CAD/CAM with
Creo Parametric

Step-by-Step Tutorial for Versions 4.0,
5.0, and 6.0

CAD/CAM with Creo Parametric

Step-by-Step Tutorial for Versions 4.0, 5.0, and 6.0

Krassimir Dotchev • Ivan Popov

University of Portsmouth, UK

World Scientific

NEW JERSEY · LONDON · SINGAPORE · BEIJING · SHANGHAI · HONG KONG · TAIPEI · CHENNAI · TOKYO

Published by

World Scientific Publishing Europe Ltd.
57 Shelton Street, Covent Garden, London WC2H 9HE
Head office: 5 Toh Tuck Link, Singapore 596224
USA office: 27 Warren Street, Suite 401-402, Hackensack, NJ 07601

Library of Congress Cataloging-in-Publication Data

Names: Dotchev, Krassimir, author. | Popov, Ivan, author.
Title: CAD/CAM with Creo Parametric : step-by-step tutorial for versions 4.0, 5.0, and 6.0 /
    Krassimir Dotchev, Ivan Popov, University of Portsmouth, UK.
Description: New Jersey : World Scientific, [2021] | Includes index.
Identifiers: LCCN 2020032971 | ISBN 9781786349330 (hardcover) |
    ISBN 9781786349453 (paperback) | ISBN 9781786349347 (ebook) |
    ISBN 9781786349354 (ebook other)
Subjects: LCSH: CAD/CAM systems. | Creo (Electronic resource) |
    Computer-aided design. | Three-dimensional imaging.
Classification: LCC TS155.6 .D67 2021 | DDC 670.285--dc23
LC record available at https://lccn.loc.gov/2020032971

British Library Cataloguing-in-Publication Data
A catalogue record for this book is available from the British Library.

For any available supplementary material, please visit
https://www.worldscientific.com/worldscibooks/10.1142/Q0274#t=suppl

Desk Editors: Ramya Gangadharan/Michael Beale/Shi Ying Koe

Typeset by Stallion Press
Email: enquiries@stallionpress.com

*The authors would like to dedicate this book
to their loving families.*

Preface

The main purpose of this textbook is to introduce a university student or practicing engineer to the basics of CAD/CAM modelling using Creo Parametric™ (Creo) software.

It would take thousands of pages to write a book about all commands and tools and address all applications of this very powerful and complex software. That is not the aim of the authors. The material in the book is meant to enable the reader to make a quick start in learning how to use a complex 3D CAD/CAM software such as Creo in engineering design and manufacturing. The aim is to develop an understanding of the main modelling principles and software tools as a basis for independent learning and solving more complex engineering problems.

The book consists of ten lessons covering Part and Assembly Modelling, Mould Design, NC Machining Simulation, and Engineering Drawings. Each lesson provides essential knowledge and guides the user through the process of performing a practical exercise or a task. The modelling philosophy, implementation of corresponding features, and commands in each exercise are explained and presented in a step-by-step manner.

The correct use of commands together with all necessary 'clicks and picks' are illustrated with icons from the software interface. All steps related to an exercise are numbered and supplemented with screenshots from the graphical user interface in order to facilitate the learning process. Each lesson should be performed with Creo software running simultaneously for better assimilation of the presented examples.

The book is not written as a reference manual with an exhaustive list of commands and descriptions covering all Creo applications. There are many aspects and applications that were not included because of volume limitations. Intentionally, the authors tried to keep the text within the minimum number of lessons required to learn the foundations of CAD/CAM modelling with Creo. It is hoped that the presented material will provide sufficient knowledge and guidance for the readers to learn the basics, continue mastering the software independently, and progress from basic to intermediate and advanced levels.

From experience, the authors believe that the first steps in mastering any 3D CAD/CAM modelling software are to learn well how to use the basic commands and also to understand the fundamental principles hidden behind the software interface. If these are practiced and learned well, then using more advanced commands and applications will be much faster. It would be only a matter of time to enhance the knowledge and skills and become a skilled user.

Despite the efforts in proofreading the text several times, there might be some inevitable mistakes. Any comments, suggestions and constructive criticism are welcome and can be send to the following e-mails: *Krassimir.Dotchev@port.ac.uk* and *Ivan.Popov@port.ac.uk*.

Enjoy the book!

About the Authors

Dr Krassimir Dotchev is a Senior Lecturer in Mechanical Engineering and CAD/CAM in the School of Mechanical and Design Engineering, University of Portsmouth. His academic and research interests encompass the areas of CAD/CAM in Mechanical Engineering, Product Design and Additive Manufacturing. He was a team leader in many research and industrial projects in the area of engineering design, rapid product development and manufacturing. Dr Dotchev has more than 20 years' experience with CAD/CAM and Creo™ Parametric as essential enabling technologies in engineering design. He holds a PhD and Dipl. Eng. in Mechanical Engineering. Dr Dotchev is a Chartered Engineer, a Member of the Institution of Engineering and Technology (IET) and a Fellow of the Higher Education Academy. He has published more than 50 technical papers in journals and conference proceedings.

Dr Ivan Popov is a Principal Lecturer in Manufacturing Engineering in the School of Mechanical and Design Engineering, University of Portsmouth. He has industrial experience in Engineering Design and Manufacturing. Currently, he is responsible for teaching Quality Management and CAD/CAM-related modules. His research interests are in Reverse Engineering, Quality Control and Manufacturing Engineering. Dr Popov has been involved in several industrial projects related to Reverse Engineering and CAD/CAM systems. He has many years of experience as an advanced user of Creo™ Parametric (former ProEngineer). Dr Popov holds a MEng and a PhD in Mechanical Engineering. He is a Chartered Engineer, a Member of the Institution of Engineering and Technology (IET) and a Fellow of the Higher Education Academy. He has published more than 15 journal papers and conference proceedings on Manufacturing Engineering, CAD/CAM and Reverse Engineering.

Contents

List of Figures

Chapter 3

Chapter 4

Chapter 5

Chapter 6

Chapter 7

Chapter 8

Chapter 1

Introduction to Creo Parametric

1.1 Introduction

Creo™ Parametric is one of the most powerful Computer-Aided Design (CAD), Computer-Aided Analysis and Computer-Aided Manufacture (CAM) software packages available in the world today. It is the flagship of a family of other software products, developed by PTC Corporation, for engineering design and product development, also including Creo™ Direct, Creo™ Simulate, Creo™ Layout and others. The main applications are in mechanical, product design, aerospace, construction, shipbuilding and other industries.

Creo™ Parametric (or Creo) was previously known as Pro/Engineer™ and Wildfire™. The core of the software contains a variety of tools for the creation, validation and communication of complex three-dimensional (3D) objects as parts and assemblies. In addition, there are integrated applications that associate directly with the 3D model geometry and support the development of engineering drawings, mould design, NC machine simulation, sheet metal design, piping and wiring, harness design, structural strength, thermal and CFD analyses, kinematic and dynamic analyses, feasibility and optimisation studies, and others. This long list of applications is not meant to scare the user but only to illustrate the vast scope and complexity of a modern CAD/CAM package.

The main purpose of this textbook is to explain the main CAD/CAM methodology and principles embedded in the software. The book is organised as a set of core lessons that provide a quick start and guide the reader in the process of mastering the basics of 3D CAD modelling.

The authors believe that if the reader learns the use of Creo, he/she would be able to master any CAD/CAM software with little effort and time. This is because the 3D modelling philosophy and core principles embedded in these systems, for example SolidWorks™, Autodesk Inventor™ and Siemens NX™, are very similar to those embedded in Creo.

The book provides ten lessons that will teach you how to create the Basic Solid Part and Assembly, Mould Design, NC Simulation and Drawings.

1.2 Tutorial Approach

The authors have adopted the tutorial method for teaching 3D modelling by demonstrating how to carry out real-world engineering examples from some very simple to more difficult models, thus illustrating the fundamental CAD/CAM topics. All lessons are presented in a step-by-step manner and illustrated with figures and icons from the interface in order to guide the learner through the 3D modelling process.

The user should be aware that a certain background knowledge in mechanical engineering and manufacturing is required for better understanding of the commands and their application.

The book should be used alongside Creo 4.0 and the next versions 5.0 and 6.0. Changes for the Creo™ Parametric version 5.0 and 6.0 were very minor compared with the basic commands and applications covered in this book.

All illustrations of the software interface and icons have been taken from Creo™ Parametric 4.0 user interface, and there might be some insignificant discrepancies with the other software releases.

1.3 Before Getting Started

All lessons (chapters) are intended to be used alongside a running Creo software. Please make sure that the software is installed on your personal computer (PC) or laptop before continuing to the next lesson.

Each of the lessons will take 2–5 hours to complete except lessons 4, 8 and 9, which might take a bit longer.

1.4 Using This Tutorial

Carefully read the content on each page and perform the given steps before proceeding to the next page. If necessary, repeat the lesson until you learn how to do the task without help. It is essential to master each topic well in order to progress to the next chapter.

There are several conventions used in this book:

- The picks and clicks are shown in **Bold** together with the specific graphic icon;
- 'Click' means to press and release the left mouse button, unless specified otherwise;
- Creo features, all commands, tools, specific interface tabs, groups, icons, drop-down menus, dashboard options, etc. are shown in **Bold**;
- Values or text to be entered are shown in **Bold**;
- Icons and their names are shown in line with the text;
- Names of models (files) are shown in CAPS;
- Keyboard keys are shown in CAPS.

Abbreviations used:

- Creo (PTC Creo™ Parametric);
- CAD (Computer-Aided Design);
- CAM (Computer-Aided Manufacture);
- LMB (left mouse button);
- MMB (middle mouse button or wheel);
- NC (numerical control);
- RMB (right mouse button);
- RAM (Random Access Memory).

1.5 Icons Used in This Textbook

Various icons are used throughout the textbook. Those related to the tutorials are as follows:

- Information is provided at the start of most tasks;
- Tips are provided along the way;
- Notes are provided as additional information.

1.6 Aim and Outcomes

Aim:

The main aim of this chapter is to assist Creo software installation and introduce the main user interface.

Outcomes:

At the end of this lesson, the reader should be able to:

- Understand the hardware requirements;
- Install Creo software and distinguish between <u>Commercial</u>, <u>Academic</u> and <u>Student</u> software versions;
- Launch Creo and understand the main interface;
- Set up a Working Directory, file names and conventions;
- Understand **File** menu, **Manage File** options, and different file types;
- **Open** a Creo object and manage files;
- Understand what a **Session** is and how to **Manage Session**;
- Understand mouse buttons and their actions;
- Understand the **Model** interface (<u>Part mode</u>) and feature dashboard.

1.7 Creo Installation

There are two main editions of Creo software: Commercial for companies, and Academic for universities and colleges. These are intended for either commercial or academic use of the software and therefore address two different communities.

There is also a Student edition targeting mostly secondary schools and colleges. The Student edition has limited CAD/CAM capabilities, and it does not provide all tools and applications available in the Commercial and Academic editions. In addition, it has limited interoperability. The Academic edition has the same tools and applications as the Commercial edition; however, the Academic license allows the use of Creo only for educational purposes.

1.7.1 *Hardware requirements*

The minimum computer system requirements are listed below:

- Operating system: Windows 7 or Windows 10 64-bit editions;
- Processor: Core i5 or higher;

- RAM: More than 4Gb;
- Monitor: Minimum resolution — 1280 × 1024 pixels with 32-bit colour;
- Hard or solid state disk memory available: More than 10Gb;
- Dedicated Graphics card: 1Gb minimum;
- Microsoft compatible 3 button (2 button and a wheel) mouse;
- Internet connection.

Creo installation software package has an automated SETUP.EXE executable file that guides the installation process of all main applications and defaults. There are some options (optional applications) that can be selected either during the first installation process or later if they are needed. However, the reader should make sure that there are valid licenses in place for all installed applications.

Run SETUP.EXE file with administrative privileges to start the installation process. The **PTC Installation Assistant** appears, as shown in Figure 1.1. The Introduction, Software Agreement and Licenses are all self-explanatory sections relevant to a specific user. The applications

Fig. 1.1.   Installation Assistant window.

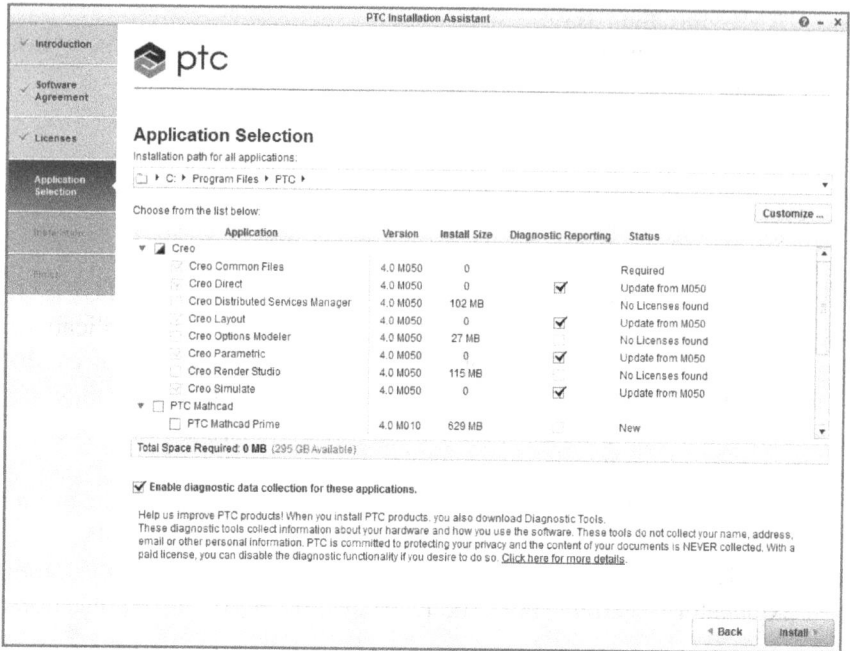

Fig. 1.2.   Application Selection window.

shown in Figure 1.2 are automatically selected based on the user license (Commercial or Academic).

By pressing the **Customize** icon, the user can access a submenu with **Options** (Figure 1.3). Make sure that **Modelcheck** and **Mold Analysis** are selected. These options are needed for the Mould Design chapter. Note that depending on the license agreement some of the applications are automatically selected (ticked) and those with no license are unavailable (dimmed) for installation. By pressing the **Install** icon (Figure 1.2), the installation starts and in the end a shortcut Creo icon is created automatically on the PC desktop.

1.8 Getting Started and Creo Interface

1.8.1 *Launching the software*

Launch Creo by LMB double-clicking on the shortcut **Creo Parametric** icon in the PC desktop. Another way to launch the software is to press on

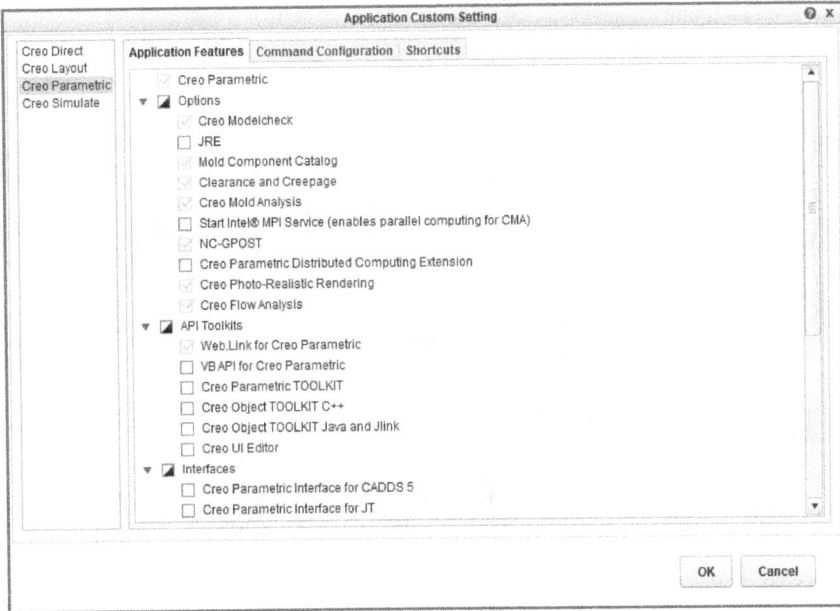

Application Custom Setting		❷ ✕

Creo Direct
Creo Layout
Creo Parametric
Creo Simulate

Application Features | Command Configuration | Shortcuts

- ☑ Creo Parametric
  - ▼ ☑ Options
    - ☑ Creo Modelcheck
    - ☐ JRE
    - ☑ Mold Component Catalog
    - ☑ Clearance and Creepage
    - ☑ Creo Mold Analysis
    - ☐ Start Intel® MPI Service (enables parallel computing for CMA)
    - ☑ NC-GPOST
    - ☐ Creo Parametric Distributed Computing Extension
    - ☑ Creo Photo-Realistic Rendering
    - ☑ Creo Flow Analysis
  - ▼ ☑ API Toolkits
    - ☑ Web.Link for Creo Parametric
    - ☐ VB API for Creo Parametric
    - ☐ Creo Parametric TOOLKIT
    - ☐ Creo Object TOOLKIT C++
    - ☐ Creo Object TOOLKIT Java and Jlink
    - ☐ Creo UI Editor
  - ▼ ☑ Interfaces
    - ☐ Creo Parametric Interface for CADDS 5
    - ☐ Creo Parametric Interface for JT

OK   Cancel

Fig. 1.3.   Application Custom Setting window.

the Windows Start icon, and then to activate **Creo Parametric** from the **PTC** applications list. The initial screen after starting the software is shown in Figure 1.4. It consists of **File** and **Home** ribbon menu options, **Folder Browser** tab, and **Favourites** tab to run a Web browser. Close the **Resource Centre** window and the Web browser link http://ptc.partcommunity.com/. The main interface at Start Up looks as shown in Figure 1.5.

The main areas and functions of this interface are as follows:

The **Folder Browser** tab (Figure 1.5), located on the left of the screen next to the **Model Tree** tab, lists the folders on the computer or network. Browse the folders and view their contents in the **Folder Browser**.

The **Favourites** tab is a multi-functional Web browser embedded in Creo. It displays models and tutorials from PTC.com and other websites. A preview window appears in the centre of the screen after the launch.

The **Folder Browser** icons (⊞ 🌐) are located at the bottom left corner of the main window. Click (LMB) on the icons to either show or hide the current folder and Web browsers.

The **Graphics** area is a large grey area where the 3D model geometry is visible when a model is opened (Figure 1.8). This is the main area

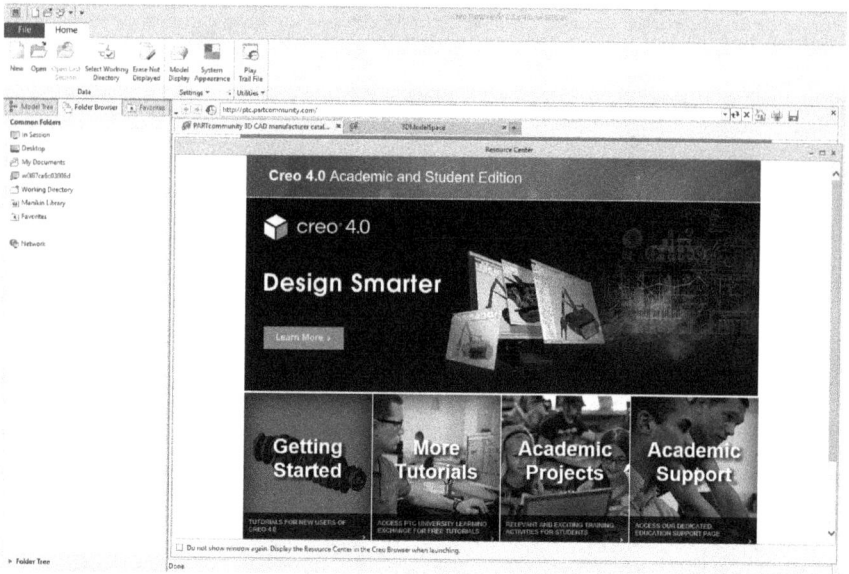

Fig. 1.4.   Creo 4.0 upon starting up.

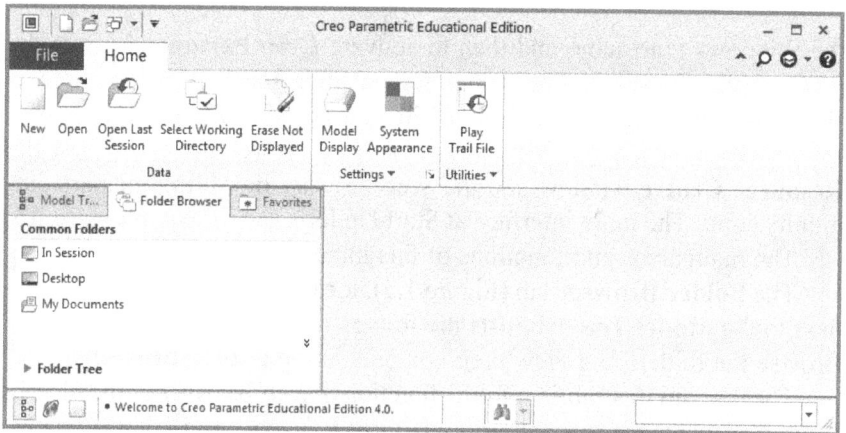

Fig. 1.5.   Home menu at Start Up.

where the model geometry is developed and displayed. The user can rotate, move and zoom the model.

The **File** and **Home** ribbon interfaces are described in the following sections.

1.8.2 *Select working directory*

The Working Directory is a directory or a folder in the PC disc memory (hard drive, solid state disc memory, USB memory stick, network, etc.) where all models (objects) belonging to a project are stored. The user can set any existing folder with Read/Write permissions as a Working Directory. It is very important to do this immediately after starting Creo or switching to a new project. If a Working Directory is not selected, the software will save or open an object from the default directory (usually in C: drive). For example, an assembly model containing several parts will not open those models that are located outside the Working Directory. The missing part name will appear in red in the assembly **Model Tree**.

Rule number one is to set up a Working Directory immediately after starting the software! This ensures that all existing and new models (files) are opened or saved in one place.

Click on **Select Working Directory** icon (⌨) from the **Home** menu at Start Up (Figure 1.5) and set up a Working Directory. Alternatively, select **File > Manage Session > Select Working Directory**.

1.8.3 *File menu options*

File menu options appear when selecting the **File** tab from the menu bar as shown in Figure 1.6. Some commands might be dimmed at Start Up.

The main commands are:

- **New:**

  Click on **New** (New), (or **File > New**) to create a new model.
- **Open:**

  Click on **Open** (Open), (or **File > Open**) to open a model.
  New and **Open** commands create a new model or open an existing model from the Working Directory. They can be activated directly from the **Home** menu after Start Up (Figure 1.5).
- **Save:**

  Click on **Save** icon (Save), (or **File > Save**) to save the model in the current Working Directory, thus creating another version of the object.

  Creo does not overwrite the existing file after **Save**. It creates a new file version containing the latest modifications. This is explained in the following sections.

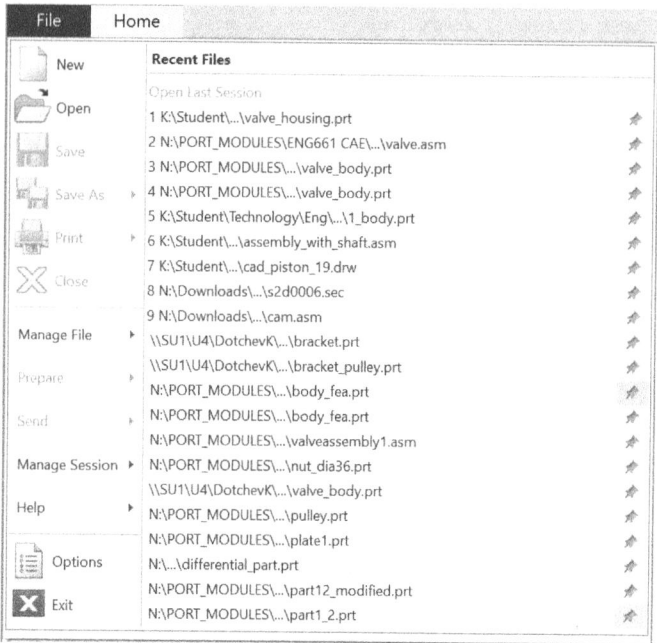

Fig. 1.6.   File menu.

- **Save As:**

  Save As (⊞ Save As ▸) icon a sub-menu with three commands:
  - **Save a Copy:** This command saves the current model with a new name. Also, it can save the model in a different file format that can be used to transfer a Creo model data to another CAD software. The most used file formats are: Stereolithography (*.STL) for 3D printing, STEP and IGES for transferring the 3D geometry into another CAD/CAM system.
  - **Save a Backup:** This command backs up an object to the current directory.
  - **Mirror Part:** This command creates a part as a mirrored copy of the current model.
- **Print:**

  Print (⊞ Print ▸) icon has a sub-menu with several commands to print the object from the active window to a printer as either a rendered image from the screen, a 2D drawing, a drawing in PDF format or a tessellated model in STL Stereolithography format.

- **Close:**
  Click on **Close** icon (⊗ ᶜˡᵒˢᵉ) to close the active window. Note that the object is still in the computer RAM, which is called **Session**. If the current model is closed accidentally, it can be opened back from the **Session** and then saved.

1.8.4 *Manage file*

Manage File (Figure 1.6) has a sub-menu with the following commands:

- **Rename (File > Manage File > Rename):**
  This command renames the current object and its versions in the **Session** (computer RAM) and/or in the Working Directory.
- **Delete Old Versions (File > Manage File > Delete Old Versions):**
  This command deletes all versions except the latest (with the highest variant number) of the current object from the Working Directory.
- **Delete All Versions (File > Manage File > Delete All Versions):**
  THIS COMMAND DELETES ALL VERSIONS of the current object from the Working Directory (from the computer disc) and the **Session**. The software asks the user to confirm this action. One should be very careful when using this command!!!

1.8.5 *Manage session*

Session is the computer RAM allocated for all Creo-opened models (objects). Several objects (parts, assemblies, drawings, etc.) can be opened simultaneously and displayed in different windows within a single **Session**. Typically, the user works simultaneously on different models, belonging to a project, and switches from one window to another. Sometimes, a new Working Directory needs to be set up without stopping Creo. In this case, all opened objects in the **Session** need to be closed and removed from it to avoid memory conflicts. This is done with the following tools from the **Manage Session** menu (Figure 1.6):

- **Erase Current (File > Manage Session > Erase Current):**
  This command removes an object (model) from the active window.
- **Erase Not Displayed (File > Manage Session > Erase Not Displayed):**

It removes all objects from the current **Session** that are not displayed.
- **Select Working Directory:** (the same command as in Figure 1.5).
- **Object List:** Display the names of all objects in session.

The best way to start a different project located in another directory is to remove all models residing in the memory (**In Session**) and then to set up a new Working Directory as follows:

Close all windows with objects.

Click on **Erase Not Displayed** to remove them from the memory. A dialogue window asks for confirmation. Before clicking on **Yes**, make sure that all listed objects have been saved.

Click on **Select Working Directory** to set up the new project directory.

1.8.6 *File names and conventions*

An object name can consist of letters and or numbers but cannot contain spaces or special symbols. The use of underscore symbol is permitted and it can also be used to separate some sections within the name. For example, MY_FIRST_MODEL is a correct name. Also, the software does not distinguish between upper and lower case letters. These rules apply to all Creo names, including the names of files, folders, features, parameters, etc.

1.9 How to Open a Creo Object (file)?

Always open a Creo object (file) from the software interface. NEVER double-click on a Creo file from the Windows **File Explorer** to open and launch the software. To avoid mistakes, follow this procedure:

(1) Start **Creo Parametric**.
(2) Click on **File > Manage Session > Select Working Directory**.
    This command sets up a directory to **Save** a new model(s) or **Open** existing model(s).
(3) Click on **File > Open** (📂).
    This command opens a window with all Creo objects from the Working Directory. Select the model name and click on **Open**.

💡 Important: Never launch Creo more than once. If you do so, you may open the models in several RAM sessions, thus preventing the association between models, for example, an assembly opened in one session and parts opened in another.

💡 The user can open as many models as needed in a single **Session**, subject to the computer RAM capacity. Then, he/she can switch from one active model (or Active window) to another at any time using the **Windows** icon (🖥 ▾), located in the Quick Access Toolbar, at the top left hand side of the screen (Figure 1.13). Click on the little black arrow (triangle) to pull down the list of all active objects in the current **Session**.

1.10 File Management in Creo

💡 In Creo, the **Save** command creates a new file with the same name but with a higher variant number and does not overwrite the previous file. After saving a model many times, the user might notice that there are several files with the same name in the Working Directory. This is a specific feature that allows the user to retrieve an older version of the same model. The question is how to distinguish between different versions. By default, the file extensions are hidden in the Windows **File Explorer**. To reveal them, open the **File Explorer**, find the **View** tab and tick the option **File name extensions** (Windows 10) as shown in Figure 1.7. The file type and extension number are now visible.

For example, a Creo file PART_MODEL.PRT.2 has the following convention: PART_MODEL is the object name, PRT is the object type

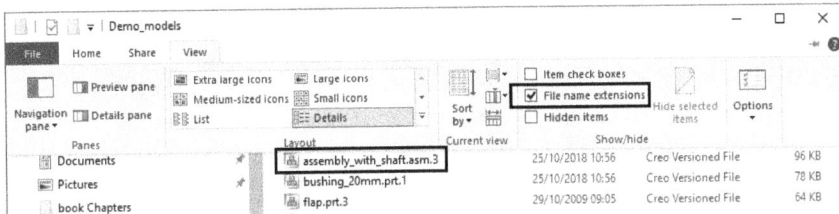

Fig. 1.7. Windows 10 explorer showing the File name extensions.

used by the software to associate it with an application (i.e. <u>Part mode</u>, or <u>Assembly mode</u>) and **2** is the model variant. After selecting **Save**, the software creates a new file in the Working Directory with the same name and type but a higher variant number **3** (i.e. PART_MODEL.PRT.3).

The command **Open** always opens the object (file) with the highest extension (variant) number.

The main object types are as follows:

- PRT — This object is created in <u>Part mode</u>. It contains a record of the model geometry as a sequence of features and their parameters.
- ASM — This object is created in <u>Assembly mode</u>. It contains the assembly model history including the names of parts, sub-assemblies and their relative locations. It may contain additional data and surface features created within the assembly. The assembly itself does not have the actual part(s) geometry. If a part model is missing from the Working Directory, then the assembly is not able to open and display the part. The part name will appear in red in the assembly Model Tree.
- DRW — This object is created in <u>Drawing mode</u>. Every drawing refers to a part or assembly model. Similar to the assembly model, the drawing does not include the model geometry and will not display it if the associated model file is missing.

The associative relationship between part, assembly and drawing is explained in Chapter 2.

1.11 Functions of the Mouse Buttons

The PC mouse is an essential input device in Creo 3D modelling. Every click of a mouse button generates an input for the software. It is important to understand and learn the functions of each mouse button when clicking on an icon (command) from the user interface or clicking on a model geometry item. In the beginning, it is difficult to remember all mouse button functions, but after some practice it will become second nature to use the correct button and key combinations. Usually, the software does not react if an incorrect button is clicked or if a wrong item selected. However, sometimes it may stop working or crash if the user repeatedly clicks a (wrong) button. Therefore, one should be careful and use the mouse buttons correctly. The main mouse button functions and combinations are explained in the following.

- **Left Mouse Button (LMB):**
  The LMB is used to: <u>click</u> and <u>initiate an action</u>, such as <u>click</u> on interface icon, tab, tool, pull down menu, open a menu and other functions. For example to <u>click</u> on Open, Save, Extrude, Rotate and other icons (commands).
  In addition, the LMB is used to <u>click</u> and <u>select</u> (pick) a geometry item as input for a command sequence. For example to <u>click</u> on datum plane, axis, point, edge, feature, etc.

Typically, the user performs a <u>click and release</u> action with the LMB to click and select (pick) an item. ALL THESE USER ACTIONS ARE DENOTED AS <u>click</u> or <u>click (LMB)</u> in the book.
  In order to make <u>multiple selections and add</u> (or remove) more than one geometrical item(s), the user must press and hold the CTRL key down (CTRL + Hold) and then continue selecting the items with the LMB.

- **Middle Mouse Button or the Wheel (MMB):**
  This is used <u>to accept</u> the current selection or input. In 2D (**Sketch**) mode, the MMB is used <u>to place dimensions</u> and also <u>to abort (stop)</u> the current command.
  MMB is also used <u>to rotate the model</u> in the Graphics area and to view it at various angles. To rotate, place the cursor on a point inside the Graphics area, press and hold the MMB (MMB + Hold) and drag the mouse slowly to rotate the model. Some computer mice have a wheel instead of a middle button. The wheel rotation zooms-In or zooms-Out the model. Rotating and zooming the model are constantly used during the modelling process.
  Another purpose of the MMB (or the wheel) is <u>to translate the model</u> on the screen. Press SHIFT and hold (SHIFT + Hold) with one finger, click the MMB, and then drag to translate the model.

- **Right Mouse Button (RMB):**
  It is used to initiate shortcut and mini menus by clicking once or by pressing and holding for a few seconds (RMB + Hold).

1.12 Creo Modelling Interface

The default, <u>Part mode</u> (part modelling) interface is shown in Figure 1.8.

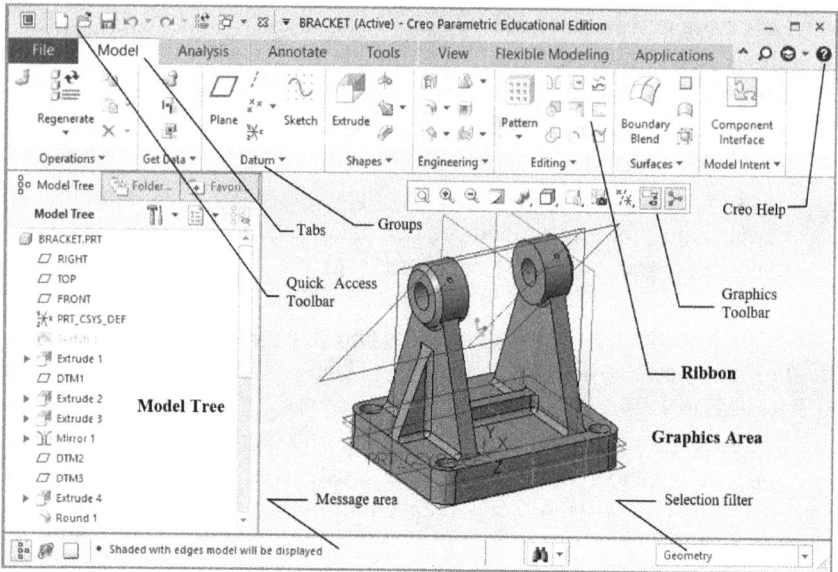

Fig.1.8. Creo part modelling interface (Part mode).

The main components of the interface are as follows:

- **Main Menu** — It is located on the top of the screen. Commands are grouped into <u>tabs</u> depending on their functions, such as **File, Model, Analysis, Annotate, Tools, View, Flexible Modelling and Applications**. Each of these tabs except **File** activates a specific <u>ribbon</u> interface. Each <u>ribbon</u> contains a number of icons, organised into <u>groups</u>, which activate commands or tools with specific actions.

The commands are context dependent, and some of them might be disabled (dimmed) if correct prerequisites are not available or if geometrical entities not selected.

The ribbon interface is active when a new or an existing Creo object is opened in the Graphics area. Figure 1.8 shows <u>Part mode</u> ribbon interface with a part opened and shown in the Graphics area. Every Creo application, i.e. part, assembly, mould design, NC assembly, and others, has a specific interface and commands.

- **Graphics Area** — This is the screen area in which the model geometry is displayed. By using the mouse, the user interacts with the

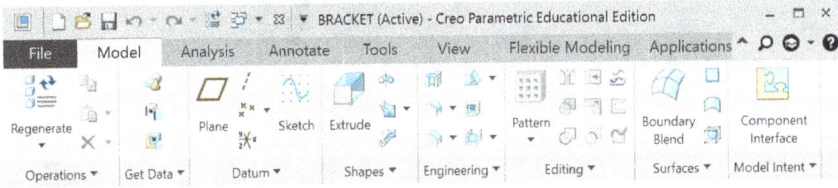

Fig. 1.9.   Model ribbon.

Fig. 1.10.   Graphics toolbar.

model, selecting geometrical entities such as plane, point, surface, edge, etc., in order to create a new model feature or modify existing model features.

- **Ribbon** — The **Model** ribbon (Figure 1.9) is a context-sensitive menu that contains the main commands (icons and drop-down menus). These are arranged in groups from left to right following the modelling workflow.

- **Message Area** — A bar at the bottom of the screen under the Model Tree and Graphics area (Figure 1.8) providing prompts and help.

Every command has a specific sequence and type of expected inputs. It is a good practice to read the messages because the system is telling what has been done and what is required next. It is recommended, especially for beginners, to read the messages during the learning process.

- **Graphics Toolbar** (Figure 1.10) — It is a small bar under the ribbon used to control the view of the model or sketch.

- **Quick Access Toolbar** — It is located on the top left of the screen. It contains the following commands: **New, Open, Save, Undo, Redo, Regenerate** and **Windows** (shown in Figure 1.11).

- The **Regenerate** command icon ( ) is frequently used in Part or Assembly mode to initiate the model recalculation after a change of a dimension or another parameter.

- **Dashboard** (Figure 1.12) — This is a submenu that will open after activation of a top-level command. For example, commands for creating Extrude, Rotate, Blend, Sweep, or other features have their dashboards. Each dashboard has command-specific tools and options for

Fig. 1.11.   Quick Access Toolbar.

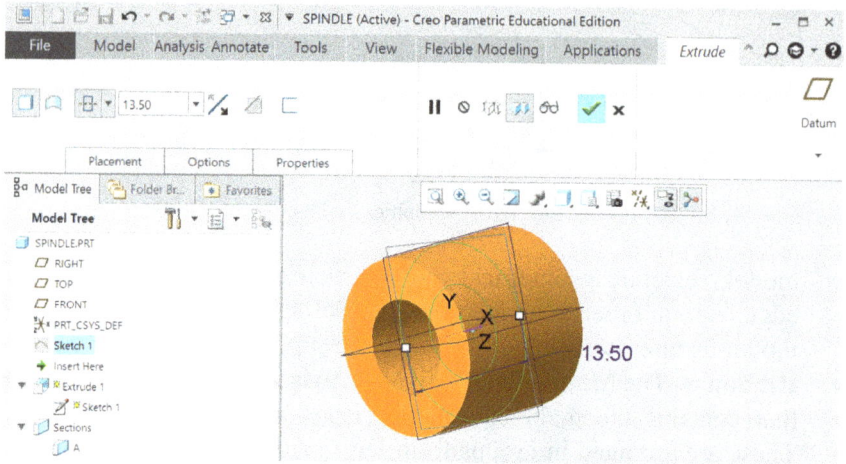

Fig. 1.12.   Extrude command dashboard.

creating variations of the activated feature. For instance, the **Extrude** command (Extrude) opens a dashboard that has options to create a protrusion (adding material) or cut (removing material). Also, the dashboard can switch to create a solid or surface feature.

Click (LMB) on **Extrude** (i.e. **Model > Shapes > Extrude**) to activate the **Extrude** dashboard (Figure 1.12). If all necessary inputs are provided into the dashboard, for example, a 2D sketch (two concentric circles) and extrude depth (Figure 1.12), then a green tick (✔) is available to save the changes and close the **Extrude** command.

Creo cannot continue if the command dashboard is still open. Click (LMB) on the **OK** green tick (✔) or **Cancel** (✘) icon on the right-hand side of the dashboard to close and continue.

• **Model Tree** (Figure 1.8) — It is a record of all features and parameters following the order of their creation. In fact, the model is a software programme containing commands and input values that Creo executes to create the model geometry. The Model Tree shows the model history. It updates every time new features are added or deleted.

- Creo **Help** — To access Creo Help Centre, click (LMB) on **Help** (❷) icon at the top right corner of the main screen (Figure 1.8).

- Use the **Search** tool (🔍) and type into the search bar to quickly find anything related to Creo, e.g. 'extrude'. The result activates the command or highlights it on the screen in real time.

- **PTC Learning Connector** (☺ ˅) — It is located next to Creo **Help**. It allows the user to log into the PTC University e-Learning database and access tutorials and videos. The user needs a <u>Login name</u> and <u>Password</u>. Most help however can be quickly sourced from the Creo Help Centre tab.

1.13 Working with Multiple Windows

The user can open several windows, each containing a different object, within the same **Session**. However, only one of these windows, the **Active window**, can be accessed for modelling. A common practice is to switch from one active window to another and work on several opened models.

This can be done by accessing the **Active window** (🗗) icon from the Quick Access Toolbar. Click on the icon, scroll down and select the model that is to become active, as shown in Figure 1.13. The current **Active**

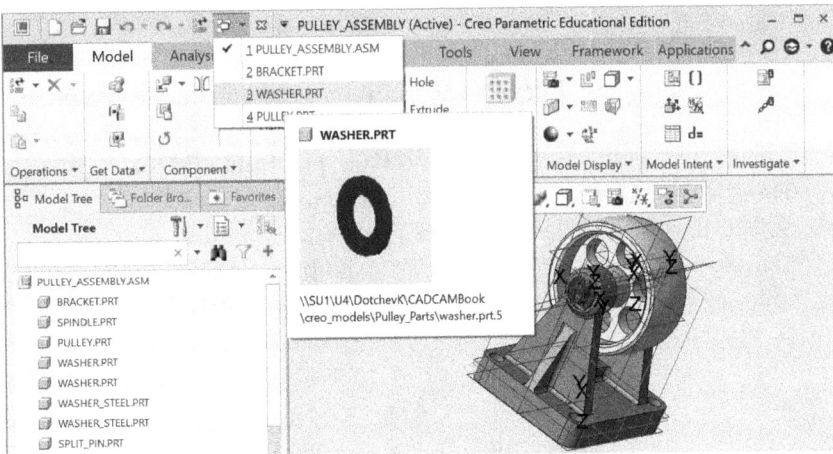

Fig. 1.13.   Active window (model) selection.

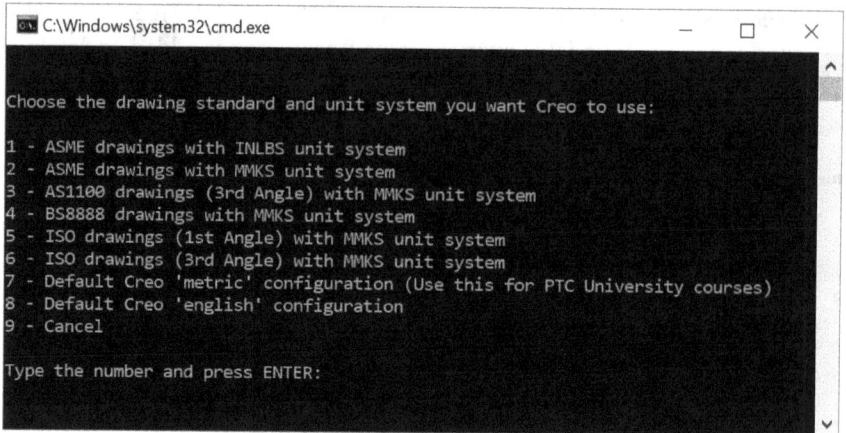

Fig. 1.14.   Unit system and drawings configuration.

window is marked with a black tick in front of the model name. After the **Active window** (object) selection, the system opens the model in the specific to the model environment, i.e. Part mode, Assembly mode, etc.

1.14 System of Units and Drawing Configuration

By default, the software will be installed and configured to the ASME Inch/Pound (Imperial) system of units. In order to reconfigure it to Metric units (mm, kg), the user should run the configuration *.BAT file as follows:

- Go to C:\Program Files\PTC\Creo 4.0\M050 (or another version)\ Common Files\creo_standards.
- Select CONFIGURE.BAT file and **Run as administrator**.
- The CMD black window (Figure 1.14) opens.
- Select the desired system from the list, type the corresponding number and press ENTER. For example, for the Metric units (mm, KG, second) select 4, 5 or 6. Select 7 for the Metric units (mm, N, second).
- Every option corresponds to a specific engineering drawing standard.

Chapter 2

Introduction to Solid Modelling
with Creo

2.1 Introduction

Solid modelling is perhaps the most important property of any modern Computer-Aided Design and Computer-Aided Manufacture (CAD/CAM) system. The solid model is a very accurate representation of the product geometry, which allows the designer not only to view the final design but also to perform engineering analyses, create engineering drawings and export NC manufacturing codes. Currently, almost all companies with businesses in engineering design and manufacture use at least one CAD/ CAM software system.

In this lesson, you will learn about the core concepts that underpin the application of solid modelling. You'll also learn, how to create a solid model using features, parameters and Parent–Child relationships.

Aim:
To introduce the key concepts embedded in a typical 3D CAD/CAM software system such as Creo.

Outcomes:
At the end of this lesson, you should be able to:

- Appreciate the advantages of solid modelling in design;
- Learn how a 3D model is created by means of a number of features;
- Understand how features are parametrically (dimensionally) driven;

- Distinguish <u>Parent–Child</u> relationships in the modelling context;
- Understand the power of associativity between models in the CAD/CAM software;
- Start learning the Part modelling interface by creating a very simple part using a Metric (mm) template;
- Start the Sketcher and create a simple 2D section;
- Create an **Extrude** feature in one or in both directions;
- Create a **Chamfer** feature.

The <u>six core concepts</u> examined in the next sections are:

- Solid and Surface modelling;
- Feature based;
- Parametric;
- <u>Parent–Child</u> relationship;
- Associative;
- Model centric.

2.2 Solid and Surface Modelling

Creo has a set of powerful applications and tools that facilitate the designer to create solid or surface models. These tools support product development from the initial concept to the final detailed design, proto-typing and manufacturing. Often, the question is what the best approach is — surface or solid modelling. Historically, surface modelling was developed earlier than solid modelling. In many cases, especially for industrial design, surface modelling provides a more flexible approach to generate complex shapes with organic curvatures. However, it takes time to learn and master the art of advanced surface modelling. In contrast, solid modelling is quite straightforward to learn, and even a beginner can create models of engineering components relatively quickly.

All modern CAD/CAM systems, including Creo, support surface and solid modelling, but also a hybrid modelling where both techniques are employed together to utilize their advantages for specific applications. In both solid and surface modelling, the 3D geometry is built as a sequence of steps or features starting with the main shapes and then creating the finer details.

The main advantage of surface modelling is the capability to generate very complex shapes consisting of surfaces with curvature in several directions such as lofts, variable blends, boundary blends, freeform surfaces, etc. These surfaces intersect, merge, offset and trim to create the final shape. In addition, surface models are more robust to modifications than solid models, and the designer can test the product appearance efficiently. However, the surface models may contain gaps between surfaces and other geometrical imperfections. Because of this, a surface model cannot be used for engineering calculation, for example calculation of mass properties, or any kind of engineering analyses.

In solid modelling, the parts are created as a sequence of 3D building blocks (or primitives) by adding or removing virtual material. As a result, the final shape is a very accurate (watertight) 3D geometry.

Solid modelling is widely used in mechanical engineering, including in aerospace, automotive and shipbuilding industries. The solid models of new products can be directly integrated with many applications within the same CAD/CAM software for verification, simulation and testing, such as assemblies, drawings, mould design, NC simulations, and strength and thermal analyses. Designers can perform accurate engineering calculations, simulate the product behaviour and improve its functionality.

Solid models have properties such as mass, volume, moment of inertia, stress, deformation, centre of gravity, surface area and other physical properties that are used in solid mechanics and dynamics analyses. Figure 2.1 shows a solid model of a piston and its mass properties. If the current dimensions are modified or new features are introduced, then the model properties will update automatically. For example, a change of the piston diameter will result in recalculation of the mass properties.

Engineering calculations such as interference/clearance checking, tolerance analysis, and kinematic (velocity, speed, acceleration) and dynamic (forces, reactions, acceleration) analyses are only possible with solid models.

It is relatively easy to learn how to create a solid model of a component with regular shape. The process is more difficult for creating complex shapes, especially for users with little or no knowledge and understanding of the main solid modelling principles. Solid models can be 'brittle' and sensitive to parametric or topological modifications, and the user should understand how the CAD/CAM system interprets the commands.

This textbook provides essential information and guidelines for a beginner to develop core skills and knowledge for efficient solid modelling with Creo software.

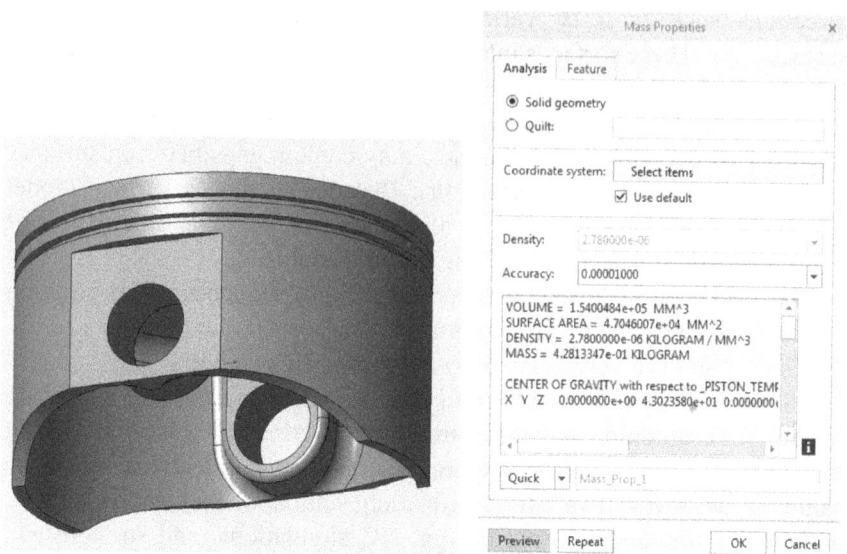

Fig. 2.1.   Solid model of a piston and its Mass Properties.

2.3 Feature-Based Modelling

Creo is a feature-based CAD/CAM modelling software. It means that all models are constructed by means of operations called features.

The type of features and their sequence in the Model Tree defines the model geometry. Every feature builds upon the previous features and can reference any of the preceding features or entities, thus enabling the design intent to be embedded in the model.

This approach is typical to all 3D CAD solid modellers and feature-based systems. The foundation of it is the Constructive Solid Geometry. Following this theory, the geometry modeller creates a complex object by applying three Boolean operations — Union, Difference and Intersection — to a sequence of defined 3D geometrical primitives. The result is the creation of a more complex shape, as shown in Figure 2.2.

Every operation (feature) that creates a shape is relatively simple. However, as they are added together, they can form quite complex parts and assemblies.

The core features are **Extrude**, **Revolve**, **Sweep** and **Blend**. To create one of these shapes, the user sketches a 2D section that is dragged along a line or a curve to form the corresponding 3D shape, as shown in Figure 2.3.

Union – Merges 2 primitives (cube +cylinder)

Difference – Subtracts one primitive from another
(cube – cylinder)

Intersection – Extracts common volume of 2 primitives

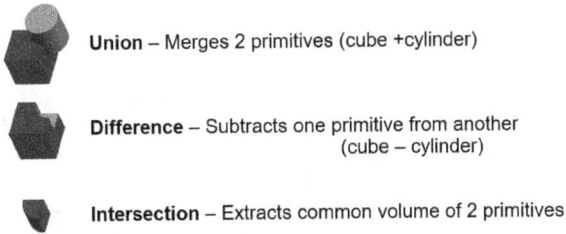

Fig. 2.2. Boolean operations with 3D primitives.

- EXTRUDE

  – Linear path of 2D Section.

- ROTATION

  – Rotation of 2D Section.

- SWEEP

  – Non-linear path of 2D section.

- BLEND
  – Connecting sequence of 2D sections.

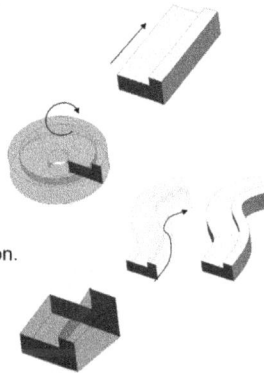

Fig. 2.3. Basic 3D modelling techniques.

Each feature can interact with the previously created features and form a Union (addition) or a Difference (subtraction) operation.

Once the main shapes are created, there are additional features (commands) such as **Round**, **Chamfer** and **Shell**, which are applied directly to reshape the geometry of any existing features (see Figure 2.4).

There are additional types of features that create supporting (datum) geometry such as **Plane**, **Axis**, **Point** and **Coordinate System**. These features are used to build a framework of references for the subsequent main features.

In addition, there are commands for feature duplication. These are **Copy, Mirror** and **Pattern**. They are directly applied to the existing features or geometry and do not require sketching on a plane.

The designer should imagine the desired geometry and then try to decompose it as a sequence of main features that either add or subtract material in order to achieve the final shape. Figure 2.4 shows an example

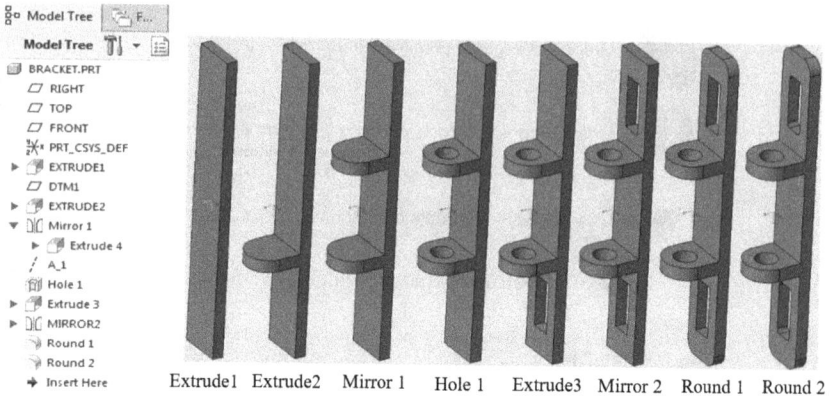

Model Tree
- BRACKET.PRT
  - RIGHT
  - TOP
  - FRONT
  - PRT_CSYS_DEF
  - ▶ EXTRUDE1
  - DTM1
  - ▶ EXTRUDE2
  - ▼ Mirror 1
    - ▶ Extrude 4
    - A_1
    - Hole 1
  - ▶ Extrude 3
  - ▶ MIRROR2
  - Round 1
  - Round 2
  - ✦ Insert Here

Extrude1   Extrude2   Mirror 1   Hole 1   Extrude3   Mirror 2   Round 1   Round 2

Fig. 2.4.   A part created as a sequence of features.

of this process applied to a component created as a sequence of several features as follows:

- The first feature, **Extrude 1**, protrudes a rectangular section to add material for the base model shape;
- **Extrude 2** uses a 2D section, sketched on a datum plane normal to the first shape, to extrude and add material;
- The third feature **Mirror 1**, mirrors the previous feature against the central datum plane;
- **Hole 1** creates a hole, subtracting material from the last two extrusions;
- **Extrude 3** (rectangular section) subtracts material from **Extrude 1**;
- **Mirror 2** creates a mirrored copy of **Extrude 3**;
- **Round 1** removes material, rounding the four external edges of **Extrude 1**;
- **Round 2** adds material, rounding the eight inside edges of **Extrude 3** and **Mirror 2**.

2.4 Parametric

All Creo models are created as sequences of features. Each feature is controlled by a set of parameters that define the feature dimensions. During the modelling process, the designer adds features (commands) to the

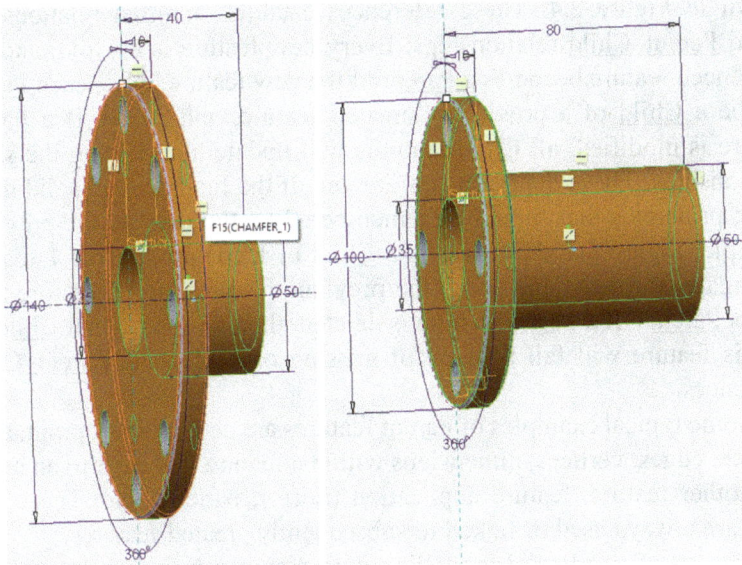

Fig. 2.5.   Flange part — before (left) and after (right) the parametric change.

model and assigns values (dimensions) to feature parameters in order to define their size and location. Each feature can be modified by editing the parameter values or redefining the feature topology. After every modification, the model should be regenerated in order to propagate the changes through all related features. Then the updated model geometry is displayed. Figure 2.5 (left) depicts a part (Flange) with **140** mm external diameter, **40** mm width and **8** axial holes. These dimensions have been modified to **100** mm external diameter, **80** mm width, and **4** holes. The new geometry after regeneration is shown in Figure 2.5 (right).

During the regeneration process, some features might fail due to collapsed geometry. In such cases, the failed features are indicated in red colour in the Model Tree. The modelling process cannot continue until the collapsed features are resolved or deleted.

2.5 Parametric and <u>Parent–Child</u> Relationships

During the 3D modelling process, every new feature creates references (links) with previously created features as demonstrated in the example shown in Figure 2.4. These references establish internal relationships called Parent–Child relationships. Every new feature is a Child, and the referenced feature becomes a Parent of the new feature. Also, each Parent can be a Child of a previously created feature, and so on. If a Parent feature is modified, all Child features will update maintaining the same relationships. These dependencies are one of the fundamental solid modelling principles that should be remembered by the user. In the previous example, Figure 2.4, **Mirror 1** is a Child of **Extrude 2**. **Hole 1** feature propagates through (intersects) the previous features, and therefore they are its Parents. If a Parent feature is deleted, then all subsequent Children of this feature will fail because of missing or inadequate Parent–Child references.

Some typical examples of Parent features are datum plane, point, axis, surface, edges, vertices, dimensions within a feature that refer to an entity of another feature, feature duplication (mirror, pattern), etc. These features are always used or linked to subsequently created features.

The art of good solid modelling is to plan carefully how to create a model as a sequence of features, anticipate potential changes, either parametric or topological, and foresee how these changes will affect the model structure and regeneration. In order to construct a resilient-to-modification model that is unlikely to fail, the user should understand well how the Parent–Child relationships work within the model and prevent eventual regeneration conflicts.

Figure 2.6 displays the Model Three of the part shown in Figure 2.5.

The external flange shape is created with **Revolve 1** feature. The holes are created with **Extrude 1** that is patterned four times. The hole centre (**Extrude 1** feature) is dimensioned to be **10** mm from the external diameter (**Revolve 1**) and because of this dimension the hole **Extrude 1** is a Child of **Revolve 1**. Thanks to this relationship, when the external diameter is modified from **140** mm to **100** mm the holes move to a smaller pitch circular diameter, maintaining the **10** mm distance from the external diameter. Alternatively, the hole centres (or the pitch circular diameter of the holes) could be dimensioned from the flange central axis. In that case the same modification of the external diameter will not affect the holes' position. However, if the external diameter becomes smaller than the pitch circular diameter, then the holes will collapse.

The feature order, dimensioning scheme, and thus the Parent–Child relationships in solid modelling are essential to the modelling process and

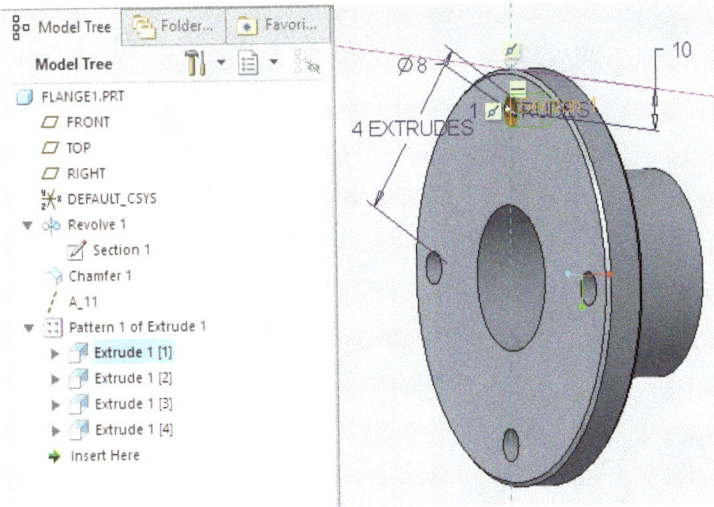

Fig. 2.6.   Editing the external diameter and hub width of the Flange model.

must be planned carefully. They define the design intent and how the model will be interpreted by the CAD system during the regeneration process. Parent–Child relationships provide powerful tools for capturing and maintaining the design intent, but they could be counterproductive in badly organized models.

2.6 Associative and Model Centric Principles

Applications such as assemblies, drawings, manufacturing models, and others refer to the 3D geometry (features) created in the part models. When the part geometry is modified, then all corresponding models and downstream applications will update automatically. This principle is known as association or associativity principle. Part models are a central source of design information and are seamlessly utilized within the Creo environment. For example, the assembly model contains the corresponding part names and does not include the actual 3D geometry. The same principle applies to all applications such as mould and manufacturing models, drawings, and others.

Typically, a part is created in Part mode independently from the assembly or drawing, which are associated with this part. The data flows

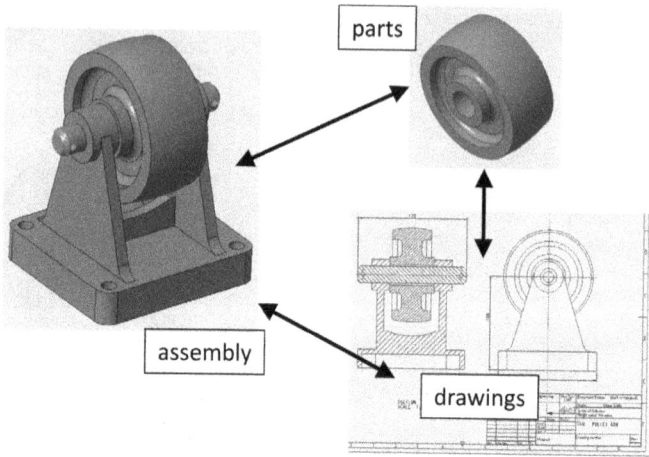

Fig. 2.7. Model associativity.

from the parts to the assembly and drawings. When a part is modified, then the assembly and drawings assume the modifications and update accordingly. Sometimes, parts can be created in <u>Assembly mode</u> (Top-down design). In this case, the data flows from the assembly to the part. Therefore, the associativity is valid in both directions — 'part to assembly' and 'assembly to part' (Figure 2.7). In relation to drawings, when part dimensions are modified, then the drawings will assume the changes. The opposite is also possible — modifications to the drawing (<u>Drawing mode</u>) can change the part geometry, although this property is rarely used. The described functionality is called bidirectional associativity (Figure 2.7).

Part models are the central source of data in the following downstream applications:

- **Assembly** — There are two assembly types: static assembly — where the parts are fully constrained and cannot move; kinematic assembly or mechanism — where parts have connections that allow them to move.
- **Drawing** — Views, orthographic projections, sectional views and annotations (dimensions, tolerances, and others) are created automatically referring to the part geometry (Figure 2.7).
- **Mold Design, NC Simulation**, and other manufacturing applications.

- Engineering Analysis applications — FEA **Structure**, **Thermal**, **Kinematic** and **Dynamic** analyses.

2.7 Creating Simple Part — Extrude and Chamfer Commands

2.7.1 *Starting Creo software*

(1) Start Creo (unless it is already running) by double-clicking on **Creo Parametric** icon on your desktop or from Windows Start icon and then finding PTC-Creo Parametric in the list.

It is a good practice to select a Working Directory at the very beginning. This is the folder where all files from a project are located.

(2) Using the Windows **File Explorer**, create a directory called C:\ USER\CREO_PRACTICE. Alternatively, the user can choose any directory with Read/Write permissions.

(3) In the Creo Start Up interface (see Figure 1.5), set up this directory as Working Directory: Click (LMB) on **File > Select Working Directory** (or **File > Manage Session > Select Working Directory**).

(4) Click (LMB) on the **New** (New) icon from the main toolbar. The **New** dialogue window (Figure 2.8) will open for the selection of the model **Type** and **Sub-type,** i.e. the application that will be initiated. By default, **Part** (Part mode) and **Solid** are selected as **Type** and **Sub-type**.

(5) A default name PRT0001 appears in the **Name** box. Click in the box to activate and then type a new name **SPINDLE**. The name should not contain spaces or special characters.

Typically, the user performs a 'click and release' action with the LMB. For simplicity, these will be denoted as click or click (LMB).

2.7.2 *New part template*

New part models are created from a template, which is an empty part containing four features — the three main (default) datum planes and a coordinate system. The default datum planes provide three orthogonal

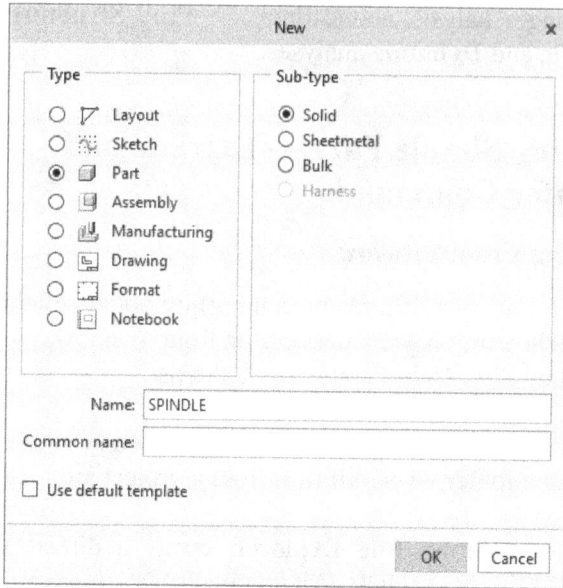

Fig. 2.8.   New Part window.

planes in the 3D space that are used as initial references for the subsequent features. Their names are **RIGHT**, **TOP** and **FRONT**.

The default template uses the American Engineering System of Units (inch, pound-force, second).

(6) *Important:* Remove the tick mark for **Use default template,** if you are not sure what is the default template, and click (LMB) on the **OK** (OK) icon to accept (Figure 2.8).

(7) The **New File Options** window (Figure 2.9) appears. Select a template with Metric Units system (metre, Newton, second), i.e. click on **mmns_part_solid**, and then **OK** (OK) to close.

With this template, all dimensions must be input in millimetres.

(8) After template selection, a new window containing the part modelling tools opens (Figure 2.10).

(9) From the Graphics toolbar (Figure 2.11), click on **Datum Display Filter** ( ) to open, then click on **Datum Axis** ( ), **Datum Points** ( ), **Coordinate Systems** ( ), and **Datum Planes** ( ) to

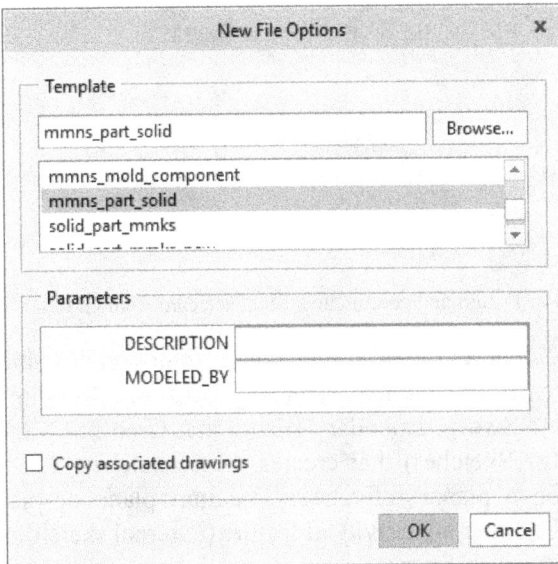

Fig. 2.9.   New File Options window.

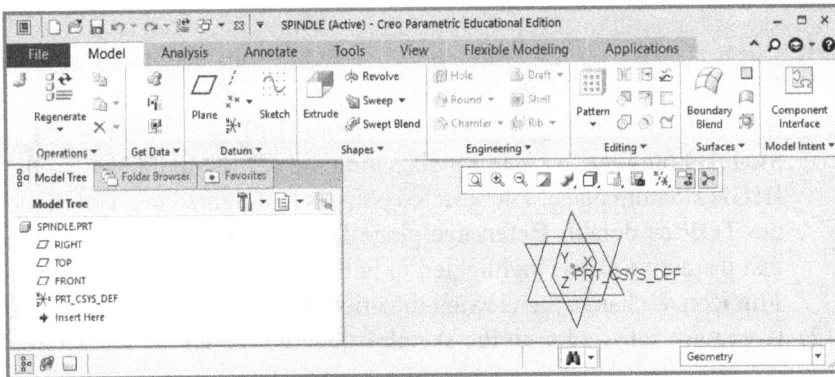

Fig. 2.10.   Part modelling window.

enable/disable their display. Disable some datums when working on complex models to better see the region of interest.

2.7.3 *Creating a sketch*

A <u>sketch</u> is a 2D feature that is drawn by means of lines, arcs, circles, curves, etc. These entities can be <u>constrained</u> (to be horizontal, vertical,

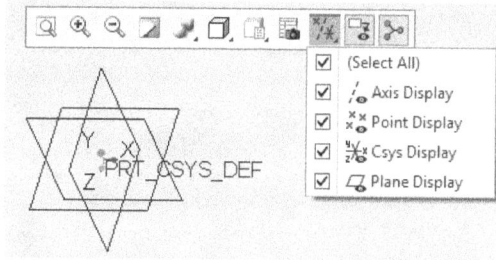

Fig. 2.11.   Datum and coordinate system selected from Graphics toolbar.

tangent, coincident, etc.) and <u>dimensioned</u>. The sketch is then used as a section to create 3D features such as Extrusion, Revolution, Sweep, Blend, Variable Sweep, Boundary Blend, etc. Creo has a module called **Sketch** tool (or Sketcher) that creates these sketches. The **Sketch** tool works only on a planar reference — datum plane or planar surface. A sketch can be either an individual feature (external sketch) or an internal sketch within a given feature.

Create a circular sketch for the main body of the model

(10) Click (LMB) on the **Sketch** ( ) icon from the main ribbon (Figure 2.12) to start the Sketcher.
(11) In the **Sketch** dialogue window (Figure 2.12) you must select a **Sketch Plane**, i.e. a plane for sketching the 2D section. Click on the **RIGHT** datum plane. The selected plane is highlighted in green. Keep the **TOP** as default **Reference** plane for **Sketch Orientation**. Note that the active slot is highlighted in light green (Figure 2.12). Use the **Flip** icon to change the viewing direction (magenta arrow) if required.
(12) If you are satisfied with the sketch orientation, click on the **Sketch** (Sketch) icon to begin. A new tab **Sketch**, containing tools for sketching (Figure 2.13), appears. To exit the **Sketch** mode, select either **OK** ( ) or **Cancel** ( ) icons. As a default, the sketching plane assumes an isometric orientation (at an angle), which is not very convenient. Click on the **Sketch View** ( ) icon, Graphics toolbar, to orient the view parallel to the sketching plane (Figure 2.14).

The Sketcher needs at least two references, one in horizontal and another in vertical directions (shown with dashed lines), in order to define

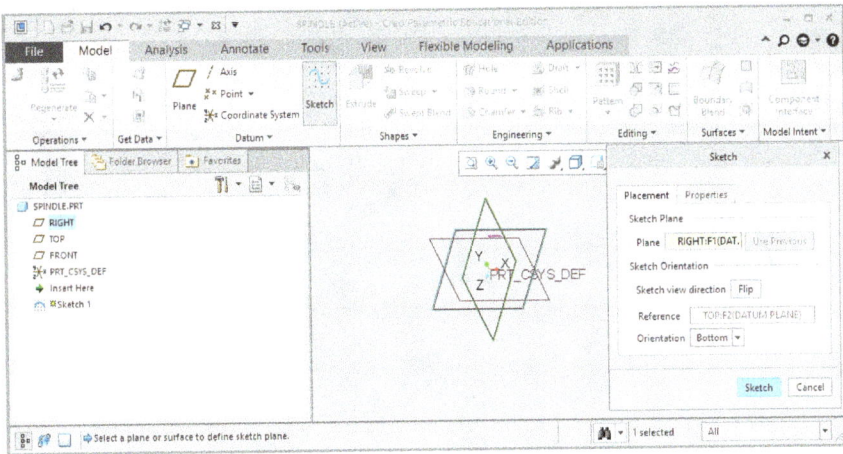

Fig. 2.12.   Selection of a plane for sketching.

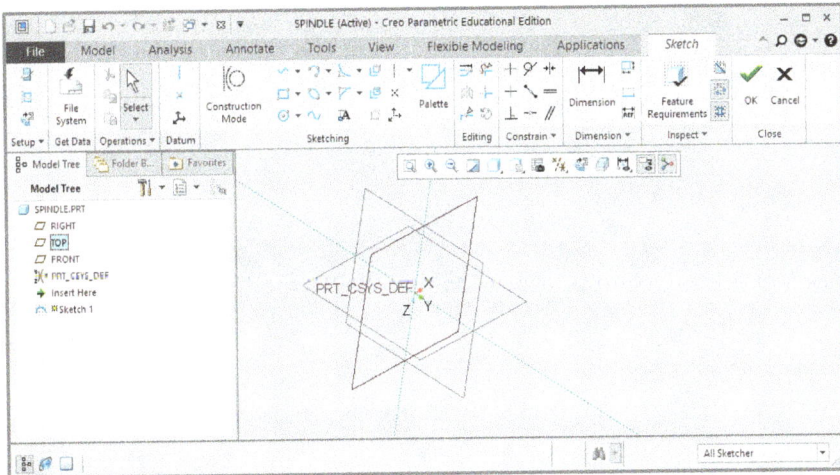

Fig. 2.13.   The start of sketch.

the location (coordinates) of the new 2D sketch. By default, these are the planes shown in profile (Figure 2.14).

💡 **Sketch Tools** mini menu. Click (RMB + Hold) on a point in the sketching area to open the **Sketch Tools** mini menu (Figure 2.15). It contains some frequently used commands such as lines, arcs, **OK** (✔) and

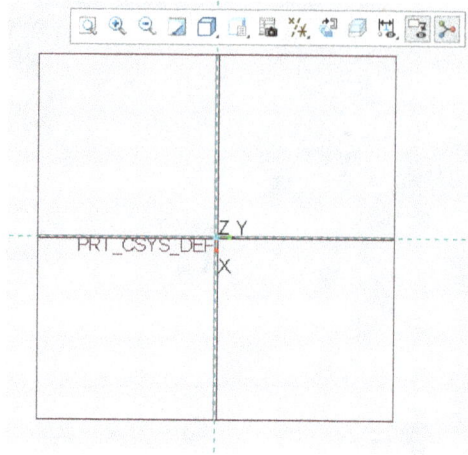

Fig. 2.14.   Sketch view, when oriented parallel to the screen, showing the two default datum planes (as dashed reference lines) and coordinate system.

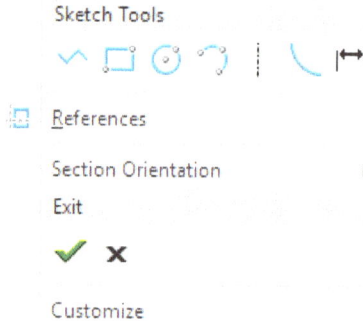

Fig. 2.15.   Sketch Tools mini menu.

Cancel (✕) icons. The full set of drawing tools, including those in the mini menu, is available in the **Sketch** ribbon. This menu is a new addition to Creo 4.0 and the later versions 5.0 and 6.0. The idea is to reduce the mouse travel and speed up the modelling process.

(13) Next, sketch a circle. Press (RMB + Hold) to activate the **Sketch Tools** mini menu (Figure 2.15) and click on the **Circle** (⊙) icon. (Alternatively, click on the **Circle** (⊙ Circle ▾) icon directly from the **Sketching** group.)

(14) Position the cursor to snap at the intersection of the two references, then 'click (LMB) and release' to place the circle centre, move the mouse outward, and 'click and release' on a point to finish the circle. Click the MMB to stop. Notice that a single dimension is created. (If the circle centre does not coincide with the reference intersection, then the circle centre will be dimensioned.)

(15) Click on **Select** (↖) from the **Sketch** tab to allow item selection.

(16) LMB double-click on the circle diameter value until it highlights, then type **15** mm and press the ENTER key to finish (Figure 2.16).

(17) Press (RMB + Hold) to activate **Sketch Tools** menu and click (LMB) on the green tick (✓) to complete and close the Sketcher. Alternatively, click on OK (✓) icon in the **Sketch** ribbon. This will bring you out of **Sketch** mode. Remember that you cannot start another feature or command until the **Sketch** ribbon is open.

(18) Click (LMB) on **Saved Orientations** (⌷) from the Graphics toolbar and select **Standard Orientation**.

(19) Click on **Datum Planes** (⌖) from the Graphics toolbar to enable their display (if not displayed).

The **Sketch** has powerful tools that are used to draw a variety of 2D entities such as **Line Chain, Circle, Arc, Ellipse, Spline Curve, Rectangle, Fillet, Chamfer, Construction Line** and other. These will be described and illustrated with examples in the next chapters.

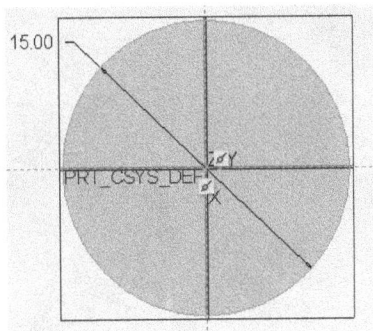

Fig. 2.16.   Circle, 15 mm diameter.

2.7.4 *Extrude feature*

The **Extrude** (Figure 2.17) is a main feature that allows for the creation of 3D geometry by dragging a sketched section to a specified distance normal to the sketching plane (e.g. a circle will make a rod when extruded).

The **Extrude** feature can <u>add material</u> to create a protrusion or <u>remove material</u> to create a cut. These options are available from the **Extrude** dashboard and will be demonstrated in this chapter.

Complex shapes can be achieved as a sequence of extrude features that add and remove material. For example, extrude a rod (cylindrical shape), and then extrude another coaxial cylinder with a smaller diameter inside and select **Remove material** option. This will create a pipe.

Create an Extruded protrusion using the previous sketch

(20) With the previous sketch selected (Figure 2.17), click on the

Extrude icon (Extrude), located in the middle of the ribbon in **Shapes** group, i.e. **Model** tab > **Shapes** group > **Extrude**. The **Extrude** dashboard opens as shown in Figure 2.18.

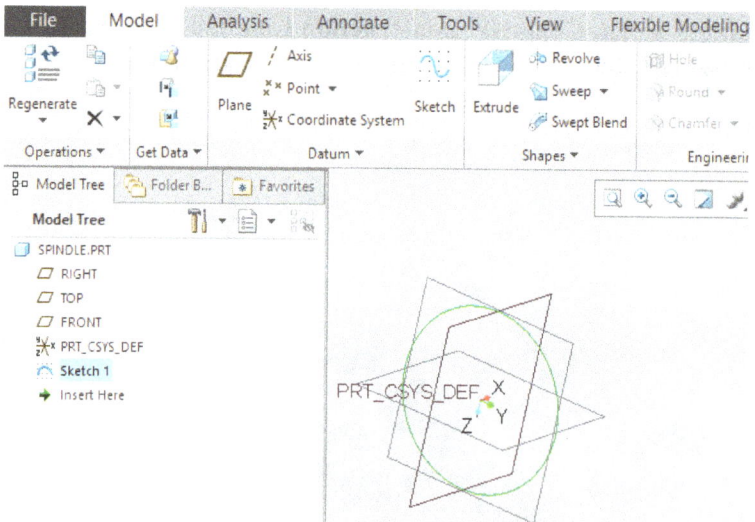

Fig. 2.17.   Sketch 1 selected (in green).

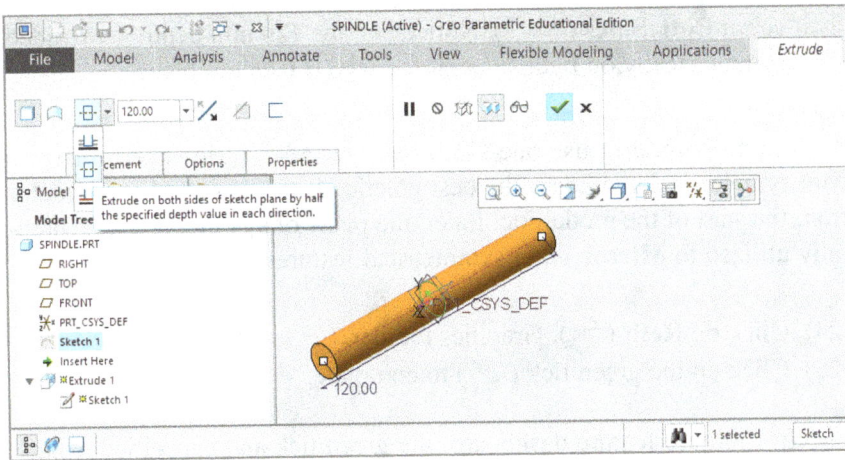

Fig. 2.18.   Extrusion of circle symmetrically in both directions.

By default, the last feature **Sketch 1** is selected and highlighted in green colour in the Graphics area. If not, then select it in the Model Tree.

Create Extrude in one direction

(21) Click on **Change Depth Direction** (⟋) in the dashboard (Figure 2.18) to flip the extrude direction. Alternatively, click on the arrow in the Graphics area to flip the direction.

(22) Click (LMB + Hold) on the depth handle (white square) and drag it to change depth dynamically. Another way to input the depth (**120**) is to type it directly in the depth area (dashboard) or double-click (LMB) on the value in the Graphics area until it highlights, type the value and press the ENTER key.

The snapping interval for drag handles is set to **1** mm (default). If a specific depth value is required, then it is more accurate to type it directly.

Create Extrude in two directions

(23) Click on the black arrow on the side of **Depth Options** (⊥ ▾) in the dashboard (Figure 2.18, top left) to open the drop-down menu, and

select **Both Sides** (⊟ ▾). Notice that the **120** mm depth is extruded symmetrically on both sides of the **RIGHT** datum plane.

💡 It is not wrong to use one-sided extrude (⊥ ▾). However, for parts with symmetrical geometry, the best practice is to promote the symmetry from the start of the modelling. Later, the plane of symmetry can be naturally utilized to **Mirror** other symmetrical features.

(24) Click on **Refit** (🔍), Graphics toolbar, to view the result.
(25) Click on the green tick (✓) to complete.

In Creo 6.0 feature dashboard, the green tick and cancel icons have **OK** and **Cancel** text added to the icons.

2.7.5 *Creating an edge chamfer at both ends of the spindle*

(26) Click on the **Edge Chamfer** (Chamfer) icon, **Model** tab > **Engineering** group. The feature dashboard opens (Figure 2.19). The **Edge Chamfer** tool needs an edge or a sequence of edges to be selected.
(27) Move the mouse pointer on the spindle left edge until highlighted in order to preselect. A label showing the edge and the feature name

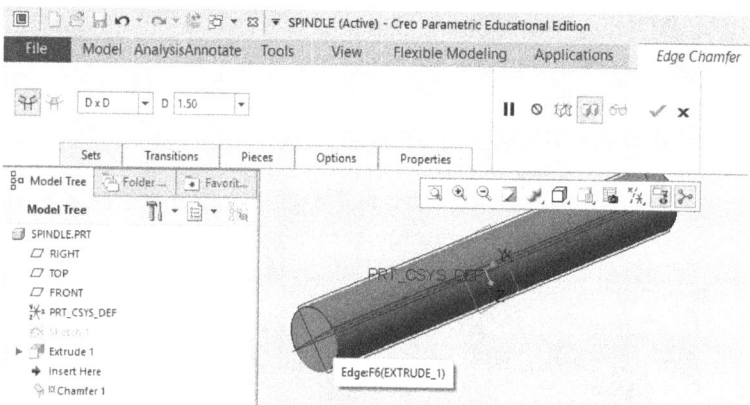

Fig. 2.19.   Chamfer dashboard.

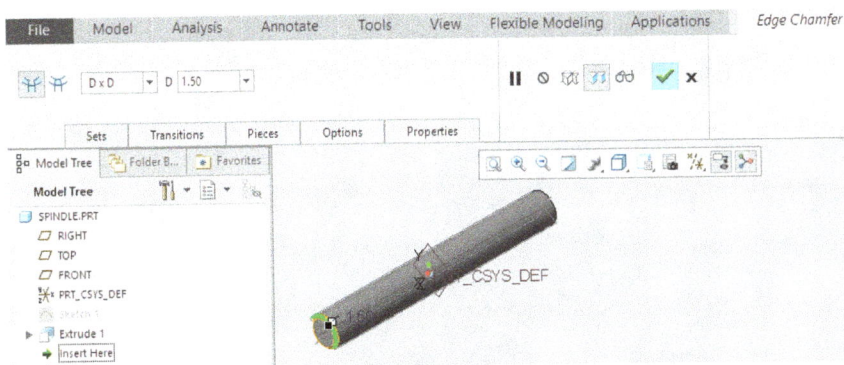

Fig. 2.20. Edge Chamfer dashboard (an edge is selected).

will appear (Figure 2.19). Click on the edge to select. If selected, the edge will highlight in green colour, as shown in Figure 2.20. The system creates a preview of the chamfer with default size.

(28) With this edge selected, press down (CTRL + Hold) and then click on the right edge to add another entity to the chamfer. Both edges should be previewed with the same chamfer size.

(29) Type **1.5** mm (chamfer dimension) directly in the **DxD** entry area of the dashboard. Alternatively click (LMB + Hold) on the chamfer handle (a little white square) to drag it and change the chamfer size.

Click on the black arrow next to the **DxD** box in the dashboard to open the pull down a menu that can switch the chamfer type as follows: **DxD** is the default 45 degree chamfer; **D1xD2** — a chamfer with unequal sides (**D1** and **D2** dimensions); and **Angle x D** — a chamfer with an **Angle** and dimension **D**.

(30) Test these options on the SPINDLE to create various chamfers.

(31) Click on the green tick (✔) in the dashboard to close.

Click on **Save** (🖫), from the Quick Access Toolbar, to save the model. The part should be saved in the Working Directory. It is a good idea to check whether this is the case and if it is not, to find out the reason.

(32) Click on **File > Close**. The Active window will close. However, the model is still in **Session** (computer RAM). To clean the memory, perform **File > Manage Session > Erase Not Displayed**.

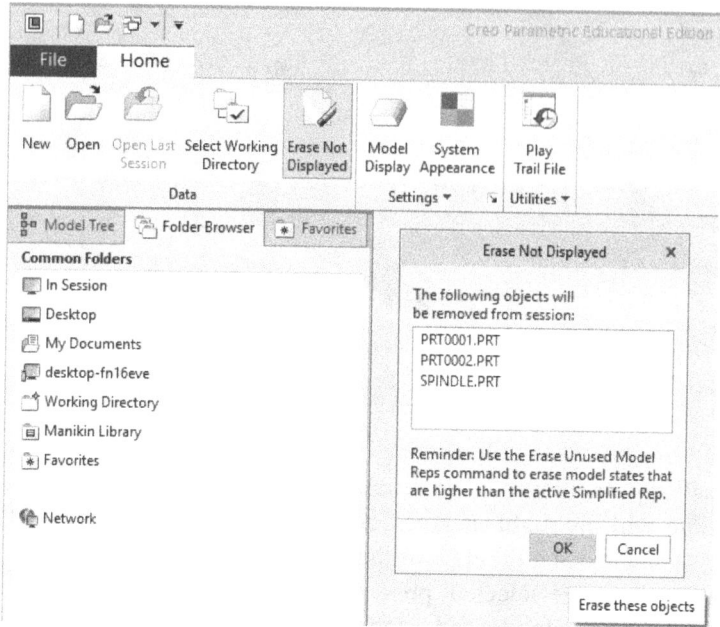

Fig. 2.21.   Erase Not Displayed window.

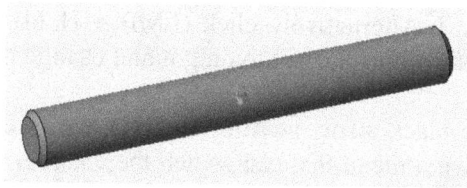

Fig. 2.22.   Spindle part.

Erase Not Displayed will clean computer RAM and the model will be lost unless saved before that with **Save** (Figure 2.21).

Congratulations! You have created your first part as shown in Figure 2.22. The SPINDLE.PRT will be developed further in the next chapters.

2.8 Exercises

Exercise 1

Run Creo. Setup a Working Directory (C:\USER\CREO_PRACTICE). Create a new part WASHER and make sure that the **mmns_part_solid** template has been selected (see Section 2.7.2). Select the **Sketch** tool and draw the washer section as two concentric circles using the dimensions shown in Figure 2.23 (left). Use **Extrude** feature and add **5** mm depth.

Exercise 2

Create a new part CAM. Select the **Sketch** tool and draw a circle **10** mm in diameter. **Extrude** the sketch by **30** mm depth. Select the cylinder flat face and draw another circle **30** mm in diameter, offset by **10** mm from the centre (for the cam section) as shown in Figure 2.23 (right). **Extrude** by **15** mm.

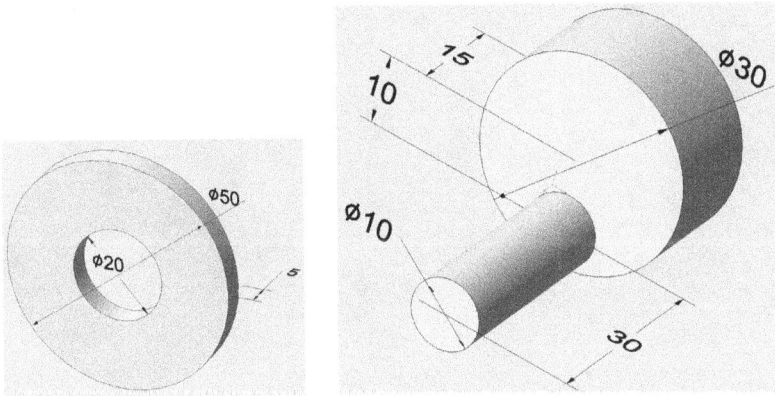

Fig. 2.23.   WASHER part (left) and CAM part (right).

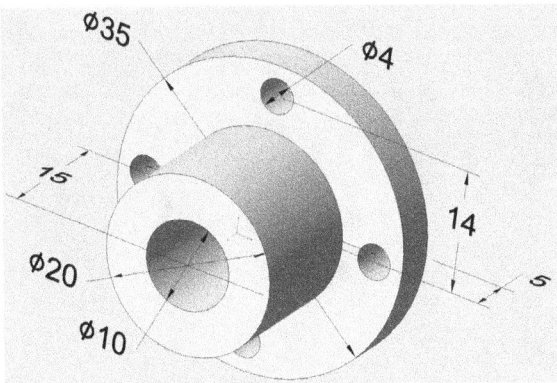

Fig. 2.24.   FLANGE part.

Exercise 3

Create a new part — FLANGE. Select the **Sketch** tool and draw two concentric circles **10** mm and **20** mm in diameter and extrude the sketch by **15** mm. Select the flat face of the extruded feature and draw another circle **35** mm in diameter. In the same section, sketch four circles **4** mm in diameter as shown in Figure 2.24 and **Extrude** by **5** mm.

Exercise 4

Use the **Chamfer** feature and create chamfers (**DxD**, **D1xD2** and **Angle x D** types) to the external edges of the parts shown in Figures 2.23 and 2.24. Gradually increase the **D** (or **D1**) value and find out what is the maximum possible value that can be applied to the chamfer.

Chapter 3

Opening, Viewing, Orienting and Editing Models

3.1 Aim and Outcomes

Aim:
In this lesson, you will learn how to open, view, change orientation and modify a model in Creo.

Outcomes:
- Learn how to use the **Folder Browser** to browse folders, preview and open models;
- Learn how to orient design models. After opening an existing model, how to use a combination of keyboard and mouse buttons to Spin, Pan and Zoom models;
- Learn how to select the model geometry and features with the mouse.
- Learn to apply colour and appearance to the models;
- Learn how the **Edit Dimensions** can modify feature parameters and change model dimensions;
- Learn how to change the topology of existing models by using **Edit Definition**, modify the sketch and change a feature option;
- Learn how to use **Suppress**, **Resume** and **Delete** commands;
- Create **Round** and **Chamfer** features;
- Lean how to use **Undo** and **Redo** commands.

3.2 Opening Models

3.2.1 *Open the SPINDLE.PRT model*

(1) Start Creo (unless already running) and set up a Working Directory, i.e. the location of the SPINDLE.PRT model in the computer disk.

ⓘ The **Select Working Directory** command sets the default folder for opening and saving models (Section 1.8.2).

(2) Open SPINDLE.PRT directly from the **Home** menu (top left corner of the screen). Click on the **Open** (⌁) icon (or **File > Open**). The **File Open** window displays the content of the Working Directory (Figure 3.1). Select the file, click on **Preview**, and then click the **Open** (OK) icon. (Another way to open the spindle model is to click on **Working Directory** (left of the screen in the **Common Folders** area under the **Folder Browser** tab), select the SPINDLE. PRT file, and then click on **Preview**. Double-click on SPINDLE.PRT to open the model.

(3) The model should appear in the Graphics area (see Figure 3.2).

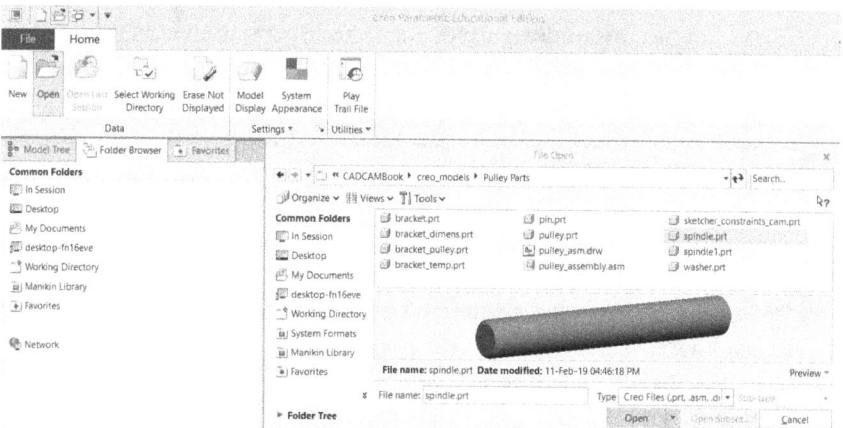

Fig. 3.1.   File Open and Preview window.

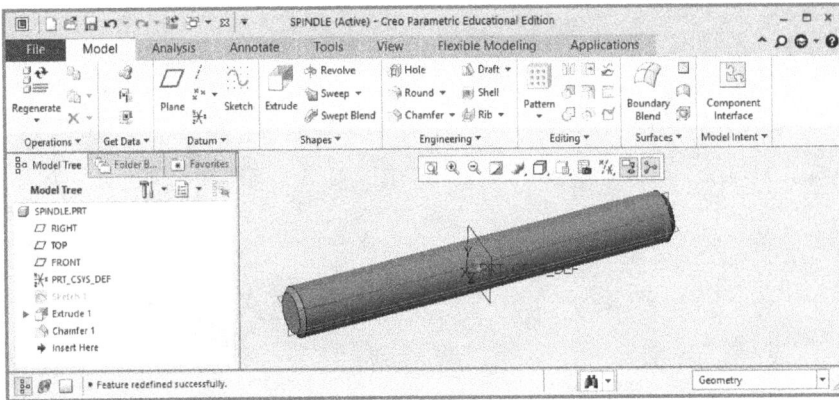

Fig. 3.2. SPINDLE model opened (Model tab).

3.3 View Control, Model Display and Orientation

ⓘ The view control, model display and orientation can be accessed from the **View** tab (Figure 3.3). Click on the **View** tab to reveal all tools arranged in the following groups:

- **Visibility** — controls the visibility of datums, curves, surfaces and parts.
- **Appearance** — applying various appearances with colour maps, scenes and photo realistic rendering.
- **Orientation** — orient mode control window, refit, zoom and allows for the user to save predefined orientations.
- **Model Display** — creating sectional views, view manager window, display style and perspective view.
- **Show** — showing or hiding datums, annotations, etc.
- **Window** — activates, closes or switches the active window.

(4) The Graphics toolbar (Figure 3.3, top of the Graphics area) duplicates frequently used **View** commands. Click on **Display Style** (⬜) and select a style for the model display, i.e. **Shading With Edges** (⬜), **Hidden Line** (⬜), and others (Figure 3.4, left).

(5) Click on **Datum Display Filters** (⬜) and deselect some or all in order to hide datum display (Axis, Point, CSys-coordinate system and

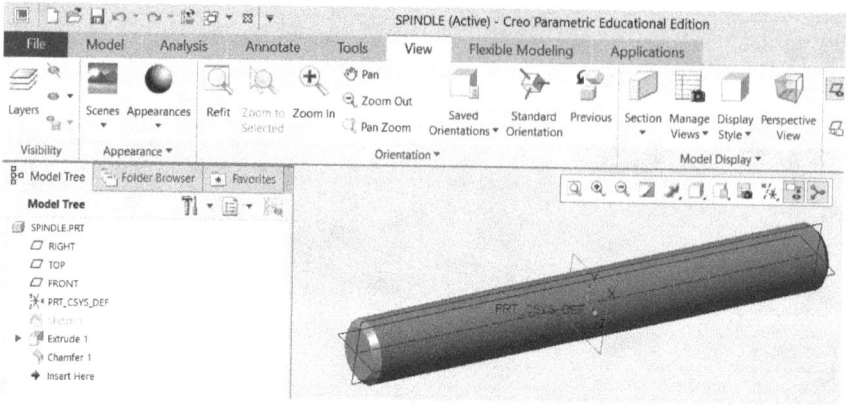

Fig. 3.3.   View tab ribbon.

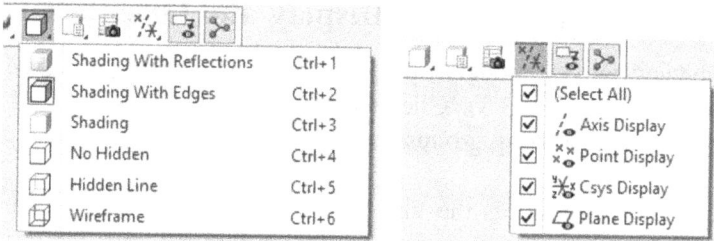

Fig. 3.4.   Graphics toolbar: Display Style (left); Datum Display Filters (right) options.

Plane Display), Figure 3.4, right. Click on the other icons from the Graphics toolbar to test their functions.

3.3.1 *Model orientation*

The most frequent model orientation commands are **Spin, Pan and Zoom**. They are activated by using a combination of the mouse and keyboard buttons. Practice the model orientation using Spin, Pan and Zoom with the SPINDLE model as follows:

Spin — Rotates the model on the screen;
Pan — Moves the model on the screen;
Zoom — Brings the model towards or away from you;

Standard Orientation — An orientation that serves as a familiar starting point for viewing the model.

(6) To **Spin** the SPINDLE model, press (MMB + Hold) and move the mouse in various directions. Notice how the model spins.

(7) From the Graphics toolbar, click on **Saved Orientations** ( ) and select **Standard Orientation**.

(8) **Zoom** in and out on the model:

- Place the mouse cursor over a desired area of the model, press (CTRL + Hold) key, then press (MMB + Hold) and move the mouse towards you to zoom in. Place the mouse cursor over a desired area of the model, press (CTRL + Hold) key, then press (MMB + Hold) and move the mouse away from you to zoom out.

The location of the cursor determines the target area for zooming.

- If the mouse has a wheel as an MMB, try zooming by rolling the mouse wheel only (without the CTRL key). Roll the mouse wheel towards you to zoom in. Roll the mouse wheel away from you to zoom out.

(9) Click on **Saved Orientations** ( ) from the Graphics toolbar and select **Standard Orientation**. Press (CTRL + D) as a shortcut to the **Standard Orientation**.

(10) **Pan** the model around the screen: Press SHIFT, then press (MMB + Hold) and move the mouse in various directions.

You can perform rotate, pan and zoom to the model in the **Preview** window when opening a model using the same mouse controls.

3.3.2 *Selecting geometry with the mouse*

Move the cursor (mouse pointer) over a feature to highlight it. If the cursor is kept still for a few seconds, then Creo will reveal the feature name and number (Figure 3.5). Click the LMB to make a selection:

- Light brown highlighting indicates that an item is pre-selected.
- Dark green highlighting indicates that an item is selected.

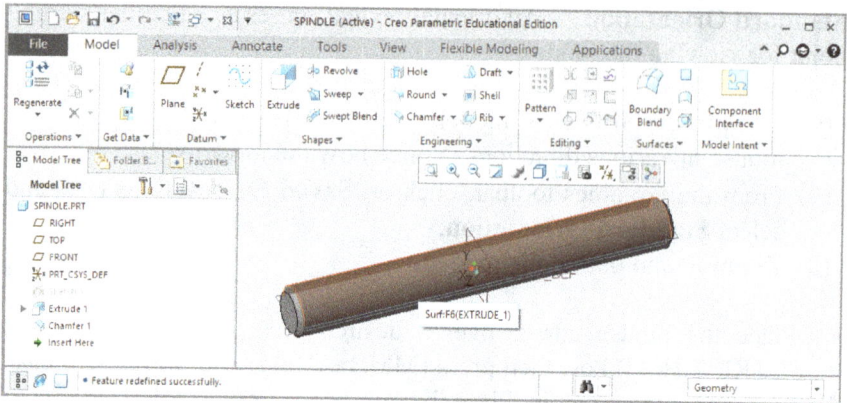

Fig. 3.5.   Highlighting geometric entity with the mouse pointer.

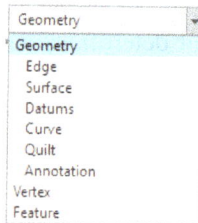

Fig. 3.6.   Geometry selection filter.

If the user selects a geometric item (surface, edge, point, etc.) then the corresponding feature in the Model Tree is also selected.

Once a feature is preselected, the edges or surfaces of the feature will highlight in light brown, and it could be selected with LMB if desired.

The 'smart' geometry selection filter, located at the bottom right corner of the screen (Figure 3.6), can be used to reduce the choice of selection. By default, it is set to **Geometry** (Geometry ▼). It means that all geometrical items will be subject to selection. The user can switch the filter to a specific entity only, i.e. to Edge, Surface, Datums, Feature, etc.

3.3.3 *Apply a colour appearance to the model*

(11) Click on **View** tab ≥ **Appearances** (drop-down menu). Select the **ptc-steel-brushed** appearance (Figure 3.7), and then select the model SPINDLE.PRT from the Model Tree and click on **OK**.

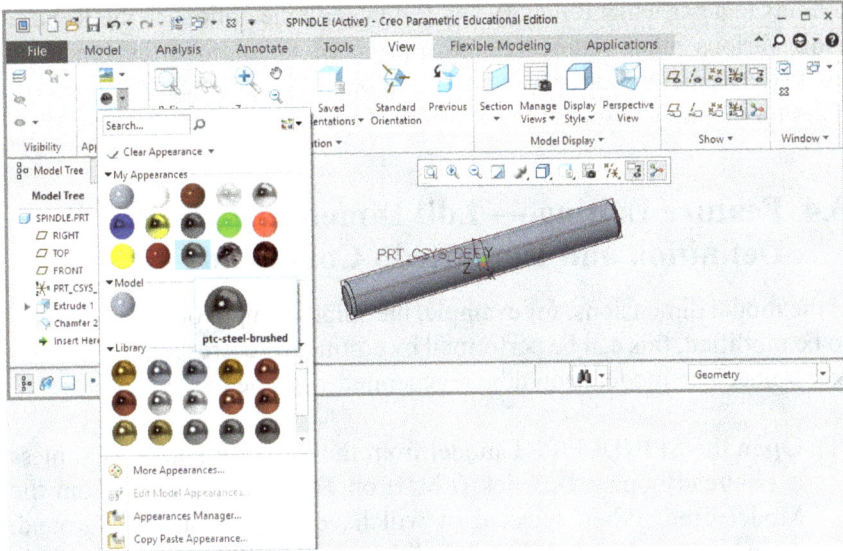

Fig. 3.7. Applying appearance to the model.

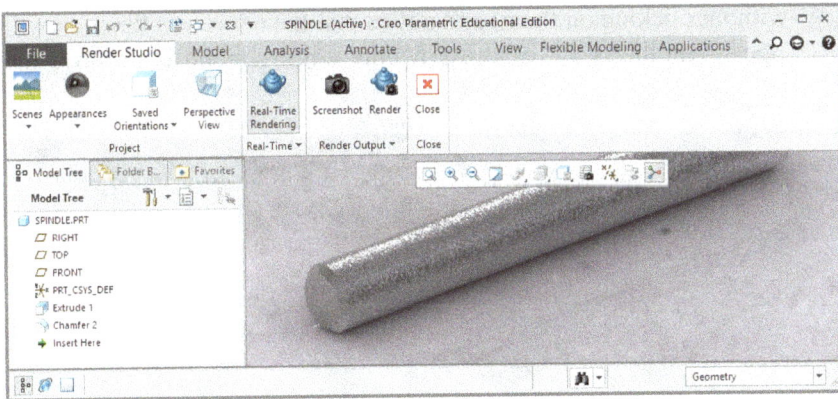

Fig. 3.8. Render Studio.

(12) Click on **Save** (🖫 ^Save) from the **main toolbar** to save the model.

(13) Click on **File ≥ Close Window**.

💡 The user can enhance the rendering quality further by using the render function in the **Applications** tab > **Render Studio** (Figure 3.8). Select the scene and press on **Real Time Rendering** to activate/deactivate. To save outputs to a particular directory and format, edit the options in the **Render** tool. Various other settings can be optimised to improve the rendering quality. The same process can be also applied to create realistic rendered animation, which essentially stitches the rendered frames together.

3.4 Feature Editing — Edit Dimensions, Edit Definition and Undo/Redo Commands

If the model dimensions, for example, the spindle length or diameter, need to be modified, this can be performed by editing the corresponding parameter value. The model should be regenerated with the new values.

(1) **Open** the SPINDLE.PRT model from the Working Directory (unless it is already opened). Click (LMB) on **Extrude 1** feature from the Model Tree. When selected, it will have a light blue background. A smart menu (a popup feature-editing menu) will appear as shown in Figure 3.9 (left). It contains the commands available for this feature. Next, move the mouse pointer slowly over the icons in the toolbar (without clicking on them) and notice that the name and function will appear. Another way to access feature editing menu is to move the

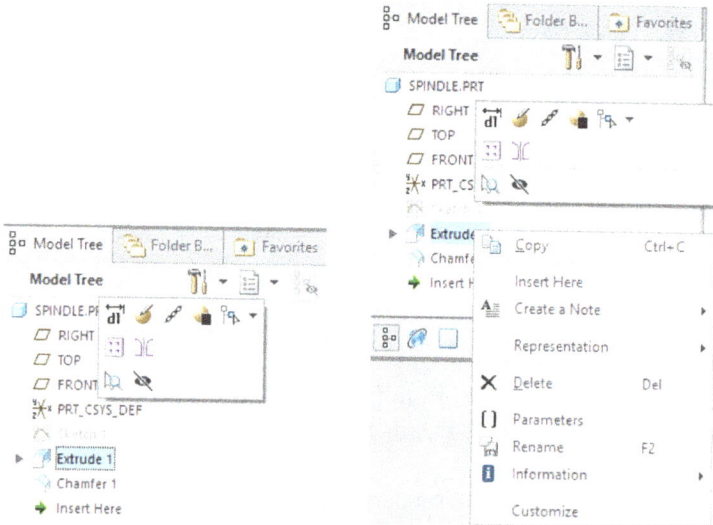

Fig. 3.9.   Feature editing — the smart menu (left) and the full pull-down one (right).

pointer on the selected feature and click RMB. In this case, a larger menu will appear next to the feature (Figure 3.9, right) containing the full list of available editing commands.

Edit Dimensions command (⊢dl⊣) allows the user to change all parameter values of the selected feature.

Edit Definition command ( ) opens the dashboard of the selected feature, allowing full access to all options to perform the desired changes.

Edit Definition command can perform topological changes to a feature. The model is rolled back to the point of the creation of the selected feature. Some changes such as modification of the sketch entities (deleting lines or creating new entities) or sketch references could eventually affect the <u>Parent–Child</u> dependency and model integrity. Sometimes, these changes are necessary to recover a failed feature. The user can create a replacement for a deleted reference. <u>Inexperienced users should be very careful when performing **Edit Definition**</u>.

3.4.1 *Edit dimensions*

(2) Select **Extrude 1** feature (unless already selected) and then click on **Edit Dimensions** (⊢dl⊣) from the mini menu. The feature parameters — the circle diameter and length — are displayed (Figure 3.10).

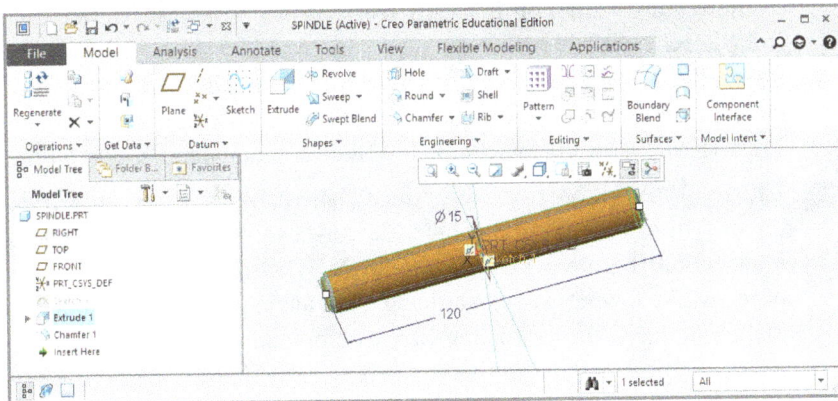

Fig. 3.10.   Editing Extrude 1 feature using Edit Dimensions command.

(3) Click and drag (LMB + Hold) the depth handle (one or two white squares at each end) to modify.

(4) Move the mouse slowly to change from **120** to **130** mm. The model shape updates dynamically.

(5) Position the mouse cursor at the yellow circle (**Sketch 1**) until it highlights, click and drag (LMB + Hold) a point from the circle to modify the circle diameter. Move the mouse slowly and notice how the diameter changes and the model updates dynamically.

(6) Double-click (LMB) on the length value, type **120** mm, and press the ENTER key to return to the initial value.

(7) Double-click on the diameter value and type **15** mm.

(8) Deselect all features with a click on the screen background.

ⓘ You can perform **Edit Dimensions** to any feature in order to display and modify its dimensions. The model should update dynamically. If it does not, then press the **Regenerate** icon ( ) from the ribbon to update the changes. A quicker way to regenerate is to press (CTRL + G).

 Always **Regenerate** if the model does not respond after **Edit Dimensions**.

3.4.2 *Edit definition*

(9) Click on **Sketch 1** from the Model Tree. A mini menu appears next to the feature. Click on the **Edit Definition** ( ) icon. Notice that the **Sketch** sub-ribbon opens (see Figure 3.11). Also, all features after **Sketch 1** are not listed in the Model Tree. They are temporally suppressed until the completion of **Sketch 1** modifications.

(10) Press (RMB + Hold) to activate the **Sketch tools** menu. Click on the **Circle** ( ) option from this menu. Alternatively, click on **Circle** directly from the ribbon. Position the cursor to snap to the references intersection (Figure 3.11) and click (LMB) to place the circle centre point. Move the mouse outward and click again to draw a smaller circle. Click the MMB to stop the **Circle** command. Double-click (LMB) on the diameter value to edit, type in **8** mm, and press the ENTER key to finish. Notice that after the second circle is drawn, the section is pink, which indicates a correct closed section, see Figure 3.12.

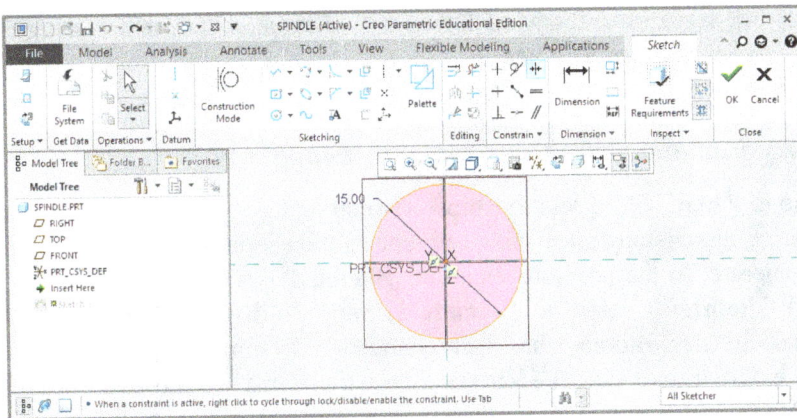

Fig. 3.11.   Edit Definition of Sketch 1.

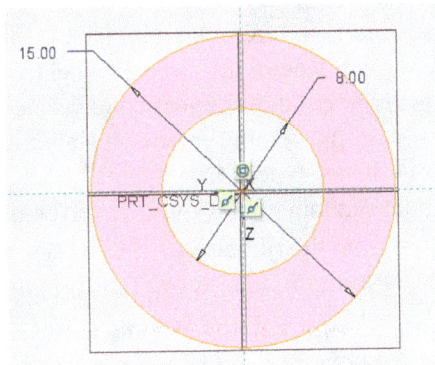

Fig. 3.12.   A second circle of 8 mm diameter.

Important! In solid modelling, the sketched section must form a closed unambiguous contour. Intersecting and/or duplicating entities or contours are not allowed. The **Sketch** tool cannot be closed until these problems are resolved. The pink colour indicates a correct closed section for solid modelling. If the section is open, then the system will automatically create a <u>Surface feature instead of a Solid feature</u>. The next solid modelling feature applied to a surface feature will fail. To correct this problem, the user need to repair the section using **Edit Definition**.

(11) Click on the **OK** (✓) icon from the ribbon (or the green tick (✓) in **Sketch tools** menu) to close the Sketcher.

(12) All features in the Model Tree will eventually regenerate using the redefined feature **Sketch 1**.

3.4.3 *Edit dimensions and undo the changes*

Due to <u>Parent–Child</u> relationships between features, if there are any conflicting dimensions or missing references, then some features may fail to regenerate. In the previous example, **Sketch 1** is a <u>Parent</u> of **Extrude 1** and **Chamfer 1** refers to **Extrude 1**. Suppose that the diameter of the inner circle in **Sketch 1** has been changed to **14** mm (**1** mm smaller than the external diameter of **15** mm), then **Chamfer 1** will collapse because its value of **1.5** mm is larger than the remaining thickness of **0.5** mm.

(13) To demonstrate this, select **Sketch 1** from the Model Tree and click on **Edit Dimensions** (⊢→⊣) from the mini menu. All **Sketch 1** dimensions are now shown. Rotate the mouse wheel (MMB) to zoom-out the model in order to enlarge the section and its dimensions.

(14) Move the cursor on the **8** mm dimension (inner diameter), double-click on it to activate, type **14**, and press ENTER to finish. The model will update automatically with an error message. Click on **Regenerate** ( ) from the ribbon (or CTRL+G) to update. The failed feature(s) appears in red in the Model Tree (Figure 3.13).

(15) Click on **Undo** ( ) from the ribbon to reverse the last **Edit Dimensions** and recover the model.

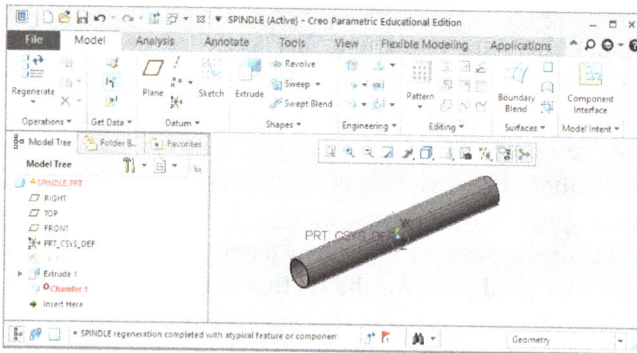

Fig. 3.13.   Failed regeneration.

Press (CTRL + Z) key combination as a shortcut to **Undo** (↺). Use also **Redo** (↻) option to revert the **Undo**.

3.5 Editing Models — Suppress and Resume Commands

The **Suppress** command temporarily removes a feature(s) from the Model Tree and from the regeneration process. If a suppressed feature is a reference (a <u>Parent</u>) to another feature, then the last will fail to regenerate properly. A suppressed feature is not deleted, and its definition remains in the model record. It can be restored with the **Resume** command.

(16) By default, the names of suppressed features are not visible in the Model Tree. They can be revealed as follows: go to **Model Tree > Settings > Tree Filters** (Figure 3.14, left) and open **Model Tree Items** dialogue window (Figure 3.14, right). Tick the **Suppressed objects** item and **OK** to close. All suppressed features are denoted by a small black square in the Model Tree. **Suppress** and then **Resume** some features from the model tree and then **Regenerate** the model.

In Creo 6.0, all suppressed features are visible by default.

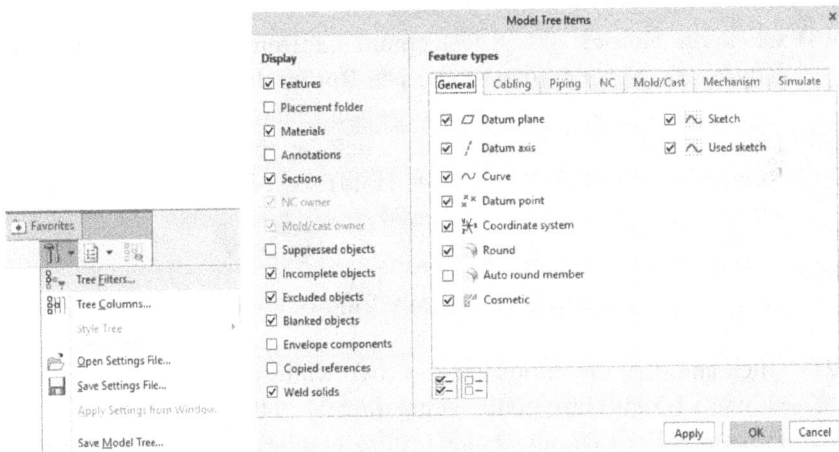

Fig. 3.14.   Tree Filters activation (left) and Model Tree Items (right).

3.6 Delete Command

The **Delete** command (see the list of commands in Figure 3.9, right) removes the selected feature(s) permanently from the Model Tree. The user can **Undo** the **Delete** command. However, after selecting to **Save** the model, all deleted features will be removed permanently. The feature(s) to be deleted can be selected either from the Graphics area or from the Model Tree.

(17) Select (LMB) **Chamfer 1**. The selected feature is highlighted in light blue colour. Click (RMB) to open the feature menu. Next, click on **Delete**, and then click **OK** (to confirm) or **Cancel**.

A faster way to delete a feature is to select the feature, then press the DELETE key, and confirm. Try this method as an alternative.

3.7 Creating Rounds

The **Round** command creates a round by applying a radius on selected external or internal edge(s) where two or more surfaces meet. You can simply select an edge (or several edges), click on the **Round** command to activate, set the radius, and complete the feature. **Round** will remove material from the external edges or add material to the internal edges.

(18) Click on **Round** (⌒ Round) command from the **Model** ribbon, **Engineering** group, to activate. The **Round** dashboard opens.
(19) Select the left spindle edge as shown in Figure 3.15.
(20) Press (CTRL + Hold) and then click on the right edge to add. Remember, that the (CTRL + Hold) key is used for multiple selections.

Notice that the round automatically follows the tangent chain.

(21) Click and drag the radius handles (two white squares) to change the value to **1.5** mm or type the value directly in the radius slot.
(22) Press MMB to complete the feature or click the **OK** (green tick (✓)) from the dashboard (top right), as shown in Figure 3.15.

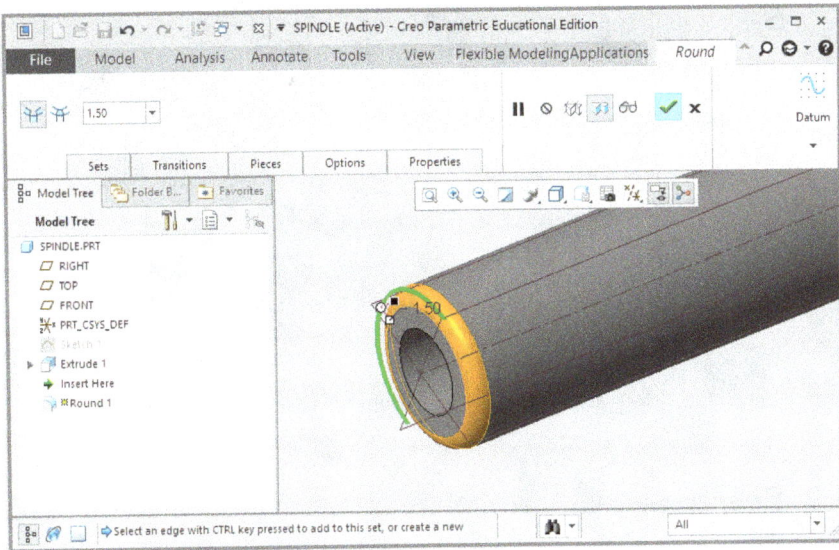

Fig. 3.15. Round feature applied on the spindle edge.

📝 Notice that **Round 1** is now the last feature in the Model Tree.

3.8 Undo/Redo Commands

💡 You can use the **Undo** (↺) and **Redo** (↻) commands to roll backward and forward the process of feature creation or correct accidental actions.

3.8.1 *Undo the created Round 1 and the deleted Chamfer 1*

(23) Click on **Undo** (↺) twice from the main toolbar to remove **Round 1** and bring back the deleted **Chamfer 1** feature, shown in Figure 3.16.

(24) Click on **Save** (🖫) icon from the Quick Access Toolbar to save the model.

(25) Click on **File > Close** (✕ Close) to close the <u>Active window</u>.

(26) The model remains in the **Session** (RAM). To remove it, click on **File >Manage Session > Erase Not Displayed** (see Chapter 1).

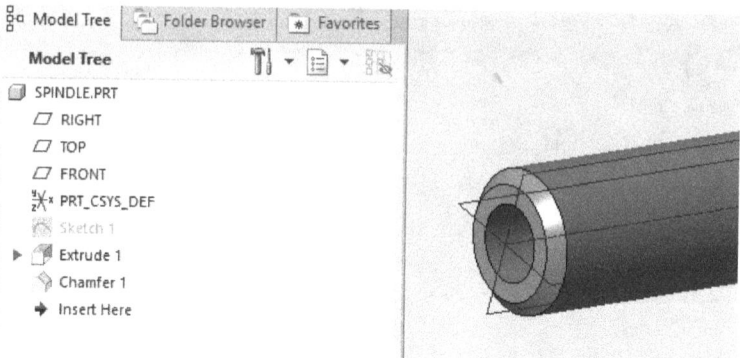

Fig. 3.16.   Undo Round 1 and return the Chamfer 1 feature.

3.9 Exercises

Exercise 1

Start Creo. Select Working Directory, C:\USER\CREO_PRACTICE. Click on **File > Open** to open the FLANGE model created in Chapter 2. Select **Sketch 1** and then click on **Edit Dimensions**. Modify the **10** mm and **20** mm diameters to **6** mm and **12** mm (double-click on each dimension, type in the new value and press the ENTER key). Select **Extrude 1** and change the depth to **25** mm. The model should update dynamically. If not, then click on **Regenerate** icon (or CTRL+G). Now modify the location of the four **4** mm diameter holes. Click on **Sketch 2**, use **Edit Dimensions** and modify the **14** mm distance to **12** mm. Note that you may have to modify the distance of each hole, i.e. four times. This is because the holes have been sketched as four separate entities. In the next chapters, you will learn how to create an **equal** constraint or to duplicate a feature in order to make a more efficient model with less parameters (dimensions).

Exercise 2

Select **Sketch 2** feature in the Model Tree of FLANGE part and then click on **Edit Definition** (✎). The Sketcher will open this feature for modifications. Delete the two opposite **4** mm circles by selecting each circle and then pressing the DELETE key. Click on **OK** to close. The model will update automatically, as shown in Figure 3.17, left.

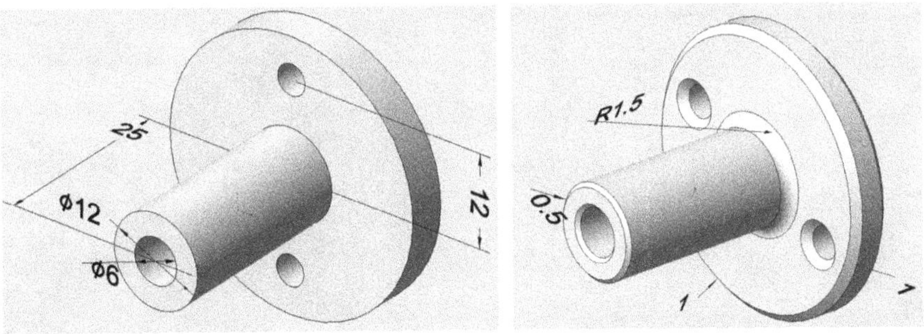

Fig. 3.17. FLANGE part: modified (left); with a round and chamfers.

Exercise 3

Create the **1.5** mm round and the chamfers on the modified FLANGE part, as shown in Figure 3.17, right.

Chapter 4

Creating Solid Part Models Using Basic Commands I

4.1 Introduction

During this lesson, the reader will continue to learn how to create part models using sketches and features and will progress to more advanced modelling techniques.

Aim:
To learn to sketch more complex sections and develop basic part modelling skills.

Outcomes:
At the end of this lesson, the reader should be able to do the following:

- Understand the design intent and plan a corresponding 3D modelling process;
- Understand the **Sketch** (Sketcher) environment and create sections using various tools available in the Sketcher;
- Inspect a sketch and identify possible drawing flows;
- Create dimensions and apply constraints to a sketch;
- Modify and redefine a sketch;
- Use **Extrude** feature with **Add** or **Remove material** options;
- Create an offset datum plane;
- Use **Mirror** command to duplicate symmetrical features;

- Create **Round** and **Chamfer** features;
- Create solid part models by means of feature sequences.

4.2 Creating Solid Part Models

In the previous chapters, the reader has already learned how to model a simple part (the Spindle) and how make basic modifications.

This lesson will demonstrate how to create the Bracket part, shown in Figure 4.1, in a step-by-step manner.

4.2.1 *Understanding the design intent and planning the 3D modelling process*

A solid model can be created in many ways using the commands and tools available in the CAD system. There is no right or wrong approach if the final geometry complies with the design specification. However, some models could contain many features and parameters that make the model difficult to interpret and modify. It is very likely that such models would fail after the modification of a single dimension. The art of a good and efficient part modelling is in creating a robust model structure with the minimum number of parameters that can be modified successfully without their geometry collapsing. With practice and experience, every designer will develop their own modelling style and rules for efficient CAD modelling, which is amenable to modifications design.

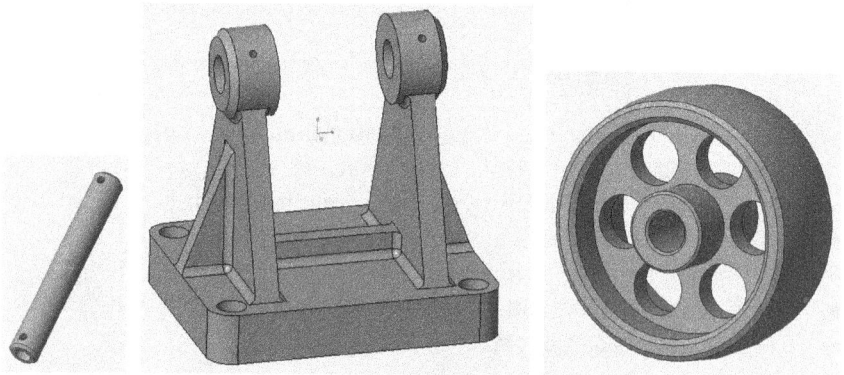

Fig. 4.1.   Basic parts models (from left to right: Spindle, Bracket, and Pulley).

A good 3D modelling workflow for a beginner would perhaps be made up of the following steps:

- Decompose the expected design into a sequence of simple main shapes (see Chapter 2, Section 2.3).
- Note those details and shapes in the design that are the same or similar. Also, notice that some elements can be symmetrical against a middle plane or can follow a certain pattern. For example, in the BRACKET part, shown in Figure 4.1, the two spindle supports and the four holes at the base are symmetrical. Creo has special commands that can duplicate, pattern, and mirror already created features.
- Start the 3D modelling and create a sequence of the corresponding features using **EXTRUDE, REVOLVE** and other basic commands.
- Add in the details and final touches such as holes, chamfers, rounds, drafts and others.
- In addition, consider how the Parent–Child relationships between features will eventually help or hinder possible modification of the model. Change the features and sequences if needed.

In the 3D modelling process, the designer should consider the following:

- The choice of specific features (commands) that create the main shapes. Different commands can result in the same geometry, but which feature is the best choice?
- Use of specific options (capabilities) within a feature (command).
- Commands for feature duplication and symmetry, i.e. Mirror, Pattern.
- The feature order in the Model Tree.
- The selection of common references.
- Sections (sketches) — selection of the geometrical entities (line, arc) and application of constraints and dimensions.

For example, when modelling the BRACKET part, it is important to note that the design is symmetrical against a middle plane. Then in the modelling, this symmetry should be embedded from the start.

4.3 Understanding the Sketcher Environment

All features that create 3D shapes involve sketches (sections and/or trajectories). Drawing correct sketches is a fundamental issue in any solid modelling process.

The **Sketch** (or Sketcher) is the module in Creo that provides all tools for drawing sketches. The Sketcher is normally activated within the <u>Part mode,</u> as demonstrated in the SPINDLE.PRT modelling, Chapter 2.

4.3.1 *Internal and external sketches*

There are two types of sketch features:

- <u>External sketch</u> — A sketch that is created as an independent 2D feature. It can be used as an external section in a 3D feature, for example as demonstrated in Chapter 2, Section 2.7.4. In this case, the sketch can be the input section in several 3D features. This approach is useful when several sketches are considered as alternatives for a 3D feature (extrude, etc.).
- <u>Internal sketch</u> — Created within a 3D feature (extrude, revolve, etc.). In this case, the sketch is embedded and used only within this feature.

4.3.2 *Capturing the design intent*

When creating solid models, it is vital to understand the design intent. For instance, symmetry, horizontal, vertical, tangent and other design considerations should be taken into account. The design intent in the Sketcher is characterized by the following:

- <u>Dimensioning scheme</u> — What dimensions are used to define the 2D shape? What are the <u>strong</u> and <u>weak</u> dimensions? Are there any sketch regenerations (problems) after changing these dimensions?
- <u>Constraint scheme</u> — Which type of constraints apply to a given 2D shape: horizontal, vertical, tangent, symmetrical, equal, or other?

4.3.3 *Sketching 2D geometry*

Learning how to create a good sketch is the basis for successful 3D modelling. Every sketch is a feature that will appear in the Model Tree.

It would be more efficient to remember the information given in the next paragraphs if Creo software is running in parallel.

(1) Start Creo, and create a new part, i.e. **New > Part**. Follow the instructions from Section 2.7, Chapter 2 until the Part mode ribbon interface (**Model** tab) opens.

(2) Click on the **Sketch** icon ( ), (**Model** tab) to start the Sketcher. Next, select a datum plane for sketching, and a reference plane for orientation (see Chapter 2, Section 2.7).

The screen displays a new sub-ribbon (**Sketch**) containing tools (icons) for drawing, editing and dimensioning. These tools are arranged in groups at the top of the screen (Figure 4.2).

The main groups and tools in the **Sketch** ribbon are as follows:

Setup group: **Sketch Setup** (Sketch Setup) tool can modify the initial setup by selecting another sketching plane without leaving the Sketcher environment. **References** (References) tool is used to add or edit the sketch references (links that 'attach' the sketched entities to the existing geometry (features)).

Datum group: Contains tools to create geometry datum **Centreline** (used in **Revolve** feature), **Point** and **Coordinate System**.

Sketching group: Contains tools for sketching entities, such as **Point, Chain Line, Circle, Arc, Rectangle, Ellipse, Spline, Fillet, Centreline**, etc.

Most entities have options that can be activated through a drop-down menu. Click on the little black arrow next to the corresponding icon to view each drop-down menu. Figure 4.3 shows the **Line** drop-down menu.

Editing group: Contains tools for editing the sketch, such as trimming (**Delete Segment, Corner**), mirroring (**Mirror**) and **Divide**.

Constrain group: Contains tools for creating constraints to the 2D entities such as: **Horizontal, Vertical, Perpendicular, Tangent, Equal, Parallel, Mid-point, Symmetric** and **Coincident**.

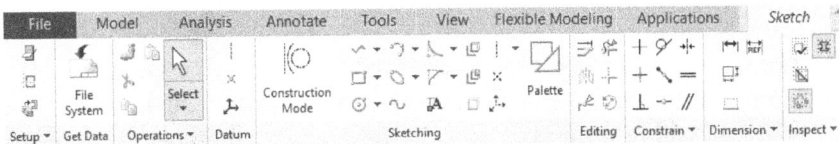

Fig. 4.2. A portion of the Sketch sub-ribbon with the main tools.

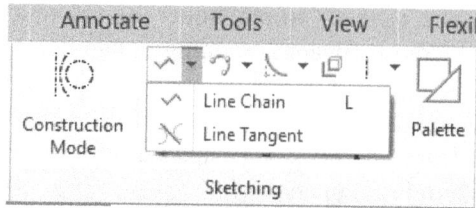

Fig. 4.3. 'Line' drop-down menu (Sketching group).

The constraints enforce specific geometrical rules to the entities. They are an important part of the sketching process that captures the design intent and adds clarity and logical conditions to the sketch. The use of constraints makes the design predictable to modifications, reduces the number of dimensions and makes the sketch, and ultimately the 3D model, robust and resilient to changes. The use of constraints is demonstrated in the next lessons.

Dimension group: Contains tools to create dimensions to the sketch.

Inspect group: Contains tools to inspect the correctness of the sketch by checking whether there is **Overlapping Geometry**, open ends (**Highlight Open Ends**) and open contours (**Shade Closed Loops**). By default, these inspections are active, but they can be turned off.

4.3.4 *The sketching group*

To start drawing a 2D entity (line, arc, circle, etc.), click on the tool icon and enter the required inputs. Once activated, the tool will continue to draw the same entity. In order to stop it, press the MMB, or click on another tool (icon).

Every tool (command) expects specific inputs in a specific order to create the entity. If a wrong input (or no input) is made by mistake, then the tool will not produce any results. In this case, the best way to resume and continue is to press the MMB to cancel, and to start again.

Always look at the message line at the bottom of the Graphics area for help. Most of the commands indicate what the expected input is.

The **Sketching** group contains the following main tools for drawing 2D geometry (sections, trajectories).

4.3.4.1 *Sketching lines*

Click on the **Line** drop-down arrow to reveal the following tools (Figure 4.3):

Line Chain creates a line or chain of lines between selected points in the Graphics area. 'Click (LMB) and release' to place a point.

Line Tangent creates a tangent line between two selected entities such as lines, arcs or circles.

4.3.4.2 *Sketching circles*

Click on the **Circle** drop-down arrow to reveal the following tools (Figure 4.4):

Center and Point creates a circle. Select a centre point first and another point in the Graphics area to indicate the circle radius.

Concentric creates a circle. Select an existing circle or arc to indicate the centre and then another point to indicate the circle radius.

3 Point tool is used to create a circle by selecting 3 points.

3 Tangent tool creates a circle that is tangent to three selected entities.

4.3.4.3 *Sketching arcs*

Arks are drawn using the tools in the **Arc** drop-down menu (Figure 4.5).

Center and Ends tool draws an arc. Select a centre point and then two endpoints in the Graphics area.

Concentric creates an arc. Select a reference circle or arc for the centre and two endpoints.

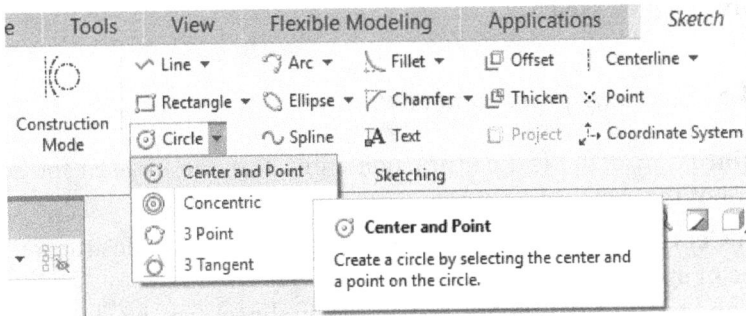

Fig. 4.4.   'Circle' drop-down menu (Sketching group).

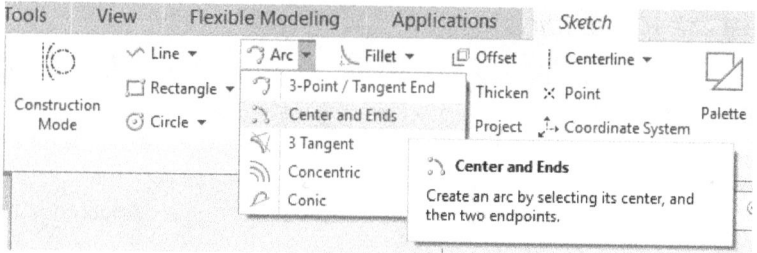

Fig. 4.5.   'Arc' drop-down menu.

3-Point / Tangent End creates an arc. Select two end points first and then a point between them.

3 Tangent tool is used to create an arc tangent to three selected entities.

💡 Move the mouse cursor on a specific tool (icon) and keep it for a couple of seconds without clicking to reveal the command Help.

4.3.4.4 *Construction geometry*

The **Construction Mode** ( ) icon (tool) triggers (on/off) the drawing of a construction entity (Line, Circle, Arc) or converts an existing geometrical entity into a construction one. A construction entity appears in the Graphics area as a dotted line or circle, which does not display in the sketch feature.

💡 Construction entities are used to create references for the sketched geometry and can be constrained and dimensioned in a similar manner. Construction geometry is very useful in creating robust and easy-to-modify complex sketches.

4.3.4.5 *Sketching points and centrelines*

A **Point** (✕ Point) is a construction entity that can be used for better control of the sketched geometry.

A **Centreline** (Centerline ▾) tool is used to create a construction line or line of symmetry in a sketch.

The construction entities (points, centrelines) can be dimensioned and/or constrained as any other sketched entity.

4.3.5 *Dimension group — dimensioning a sketch*

While drawing or after finishing a sketch, the user can initiate dimension-ing by clicking on the **Dimension** icon (⊢⇀⊣). The system will automati-cally place the minimum number of dimensions defining the sketch. These dimensions are called <u>weak</u> dimensions. However, they are not necessarily the desired dimensions. In this case, the **Dimension** (⊢⇀⊣) tool is used to manually create extra dimensions. Manually created dimensions are called <u>strong</u> dimensions. As new dimensions are created, the system automatically removes the redundant <u>weak</u> dimensions. If there is a redundant strong dimension, the system will ask the user to select the one to be removed.

The **Dimension** tool can create the following dimensions:

<u>Distance</u> — Click (LMB) on <u>two entities</u> (two parallel lines, two points, two circles, arcs, etc.) to select, move the cursor in between and click the MMB to place a dimension.

<u>Angle</u> — Click on <u>two intersecting entities</u> to select, move the cursor and click the MMB on a point in between to place a dimension.

<u>Radius</u> — Click on an arc or circle, move the cursor and then click MMB to place the radius. To switch a radius to a diameter, select the dimension, press (LMB + Hold) to open the mini menu and click the **Diameter** (⟲) option in it.

<u>Diameter</u> — Click on two points of a circle (arc), move the cursor and MMB click to place a diameter dimension.

<u>Length</u> — Click on a line to select, move the cursor and click the MMB to place the length.

4.3.6 *Modifying sketch dimensions*

Sketcher dimensions can be modified as follows:

- Double-click (LMB) on an existing dimension, type the new value and press the ENTER key;

- A sketch entity can be moved (modified) by dragging it with the mouse as follows: click on the **Select** (⊾) icon to activate, move the cursor until the entity is preselected then click and drag (LMB + Hold) the entity.

4.3.7 *Locking dimensions in the sketch*

A locked dimension cannot be modified by dragging the attached entity. It can only be changed by double-clicking on it and typing a new value.

To lock (or unlock) a dimension: Click on a dimension, then in the mini menu that appears, click on **Toggle Lock** (🔒) icon.

4.3.8 *Sketcher constraints (constrain group)*

4.3.8.1 *Intent manager*

The system automatically applies relevant constraints to the 2D entities trying to capture the designer's intent. For instance, nearly horizontal or vertical lines will attract a horizontal or vertical constraint. Also, end points that are very close will become coincident. In some cases, this is not desired. To avoid this, the designer should exaggerate, i.e. draw a line at a larger angle, or increase the gap between points.

4.3.8.2 *Applying constraints*

Constraints can be applied (added) to a sketch manually. The **Constrain** group (Figure 4.2) contains the following nine constraints:

Vertical (|) — Selected lines and points are aligned vertically;
Horizontal (⊤) — Selected lines and points are aligned horizontally;
Perpendicular (⊥) — Selected lines are made perpendicular;
Tangent (𝒪) — Selected lines, arcs, circles and splines are made tangential to each other;
Mid-point (⟍) — Connects the starting point of a new entity to the midpoint of a line or arc;

Coincident (⁻⁻) — Selected lines or points are made coincident to each other or coincident to a reference;

Symmetric (⁺⁺) — Selected points or vertices (line, arc) are made symmetric to a selected centreline;

Equal (=) — Selected dimensions are made equal; selected lines or arcs will have equal dimensions (equal length or radius);

Parallel (//) — Selected lines (and/or centrelines) are made parallel.

Figure 4.6 shows a sketch before and after applying constraints. The constraints are indicated by a small light green square, attached to the entity, with the constraint type (Figure 4.6, right). The following constraints are applied: the three lines are horizontal; one line perpendicular to the base, the two slanted lines parallel and the upper two horizontal lines equal.

4.3.8.3 *Deleting a drawing entity and constraint*

Any unwanted drawing entity (line, circle, construction line, etc.), constraint or dimension can be deleted as follows:

- Click on the **Select** ( ) icon, in the **Operations** group, move the cursor and select the item with LMB click. Press the DELETE key.
- Click on **Select** ( ) icon, press (LMB + Hold) and drag the mouse cursor to draw a rectangle around several entities to be marked simultaneously. Press the DELETE key.

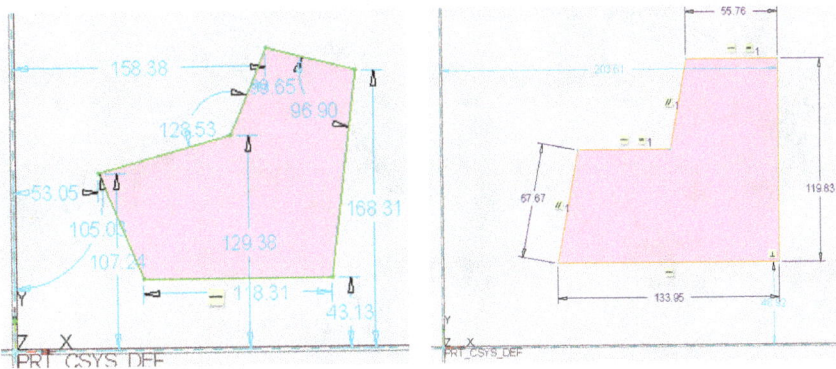

Fig. 4.6.   Sketch before (left) and after (right) applying constraints.

4.3.9 *Pick From List dialogue box*

Several constraints or entities can be located close to each other in the Graphics area, and sometimes it is impossible to select the right one with the cursor. In this case, the user can use the **Pick From List** tool as follows:

- Move the cursor to the area of selection, then press (RMB + Hold) and from the pull-down menu select **Pick From List** (Figure 4.7, left).
- The **Pick From List** dialogue box (Figure 4.7, right) will appear with a list of all relevant entities in the area of selection.
- Use the arrows to select the right entity and then click **OK** to pick the item.

4.3.10 *Resolve sketch dialogue box*

The Sketcher automatically evaluates the correctness of the drawn entities by maintaining the minimum number of dimensions and constraints that fully define the sketch. In case of a conflict such as an extra dimension or constraint, the **Resolve Sketch** dialogue box will appear, as shown in Figure 4.14. The user should delete the conflicting (redundant) dimension or constraint, or **Undo** to revert to the previous stage.

4.3.11 *Creating revolved sketches*

Revolved sketches are used for creating **Revolve** features and are drawn in the same way as any other sections. The only difference is that a revolve

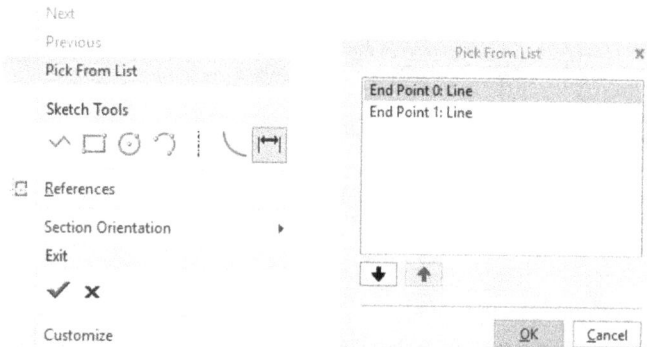

Fig. 4.7.   Pull-down menu (left) and Pick From List dialogue box (right).

sketch has a datum **Centreline (Datum** group) that indicates the axis of rotation. In addition, the section entities should not cross the Centreline.

💡 Datum Centreline is different from construction **Centreline** and is shown by a yellow colour in the sketch while the construction line has a magenta colour.

4.3.12 *Undo and Redo commands*

Use the **Undo** (↺ ▾) or **Redo** (↻ ▾) icons (top left on the screen) to return to the previous stage of the sketch creation.

4.3.13 *Editing the sketch entities (editing group)*

The **Editing** group (Figure 4.8) contains the following editing tools:

Delete Segment — Click on the icon to start the tool, then move the cursor above the entities to be trimmed and click and drag (LMB + Hold) the cursor across the entities to be dynamically trimmed or deleted. Any entity that is crossed or touched with the **Delete Segment** tool will be deleted.

Corner — Click on the icon to start the tool, then select (LMB) two intersecting entities to trim.

Mirror — Click on the icon to start, select a **Centreline** and then click on the entity to be mirrored. For multiple selections, press (CTRL + Hold) and then click on several entities.

Divide — Start the tool, select an intersecting point of entities to be divided.

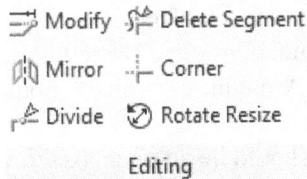

⇥ Modify ⤼ Delete Segment

🗍 Mirror ⊣ Corner

⌐ Divide ⟳ Rotate Resize

Editing

Fig. 4.8. Editing group.

Undo any (wrong) operation within the Sketcher using the **Undo** and/or **Redo** commands as described earlier.

4.3.14 *Sketcher display options in the graphics toolbar*

An additional icon ( ), **Sketcher Display Filters**, appears in the Graphics toolbar to toggle the display of dimensions, constraints, grid, vertices and locks. Dimensions and constraints are displayed by default.

4.3.15 *Inspect a sketch and identify drawing flows (inspect group)*

Several problems might occur during the process of sketching. There are some general rules and best practices that should be followed to ensure correct creation of sections for successful solid features, and these are as follows:

- Do not make very complex sections (sketches) containing too many entities and dimensions unless absolutely necessary. It is difficult to control and modify a complex sketch.
- The sketch should not contain any overlapping geometrical entities (line drawn on another line).
- Remove any small accidentally sketched entities.
- Use construction lines (also circles, arcs and points), constraints and symmetry to reduce the number of dimensions.

The following rules apply when sketching some specific solid features:

- **Extrude, Revolve, Sweep, Blend** and other <u>solid features</u> — The sketched section should form a closed loop for solid geometry. An opened sketch will create a surface.
- **Revolve** — The datum centreline for revolution should not cross the sketched geometry. A datum centreline should only be used for the **Revolve** feature.
- **Rib** — The section should be an open sketch with endpoints attached to references from the existing solid geometry.

Fig. 4.9.   Inspect tools.

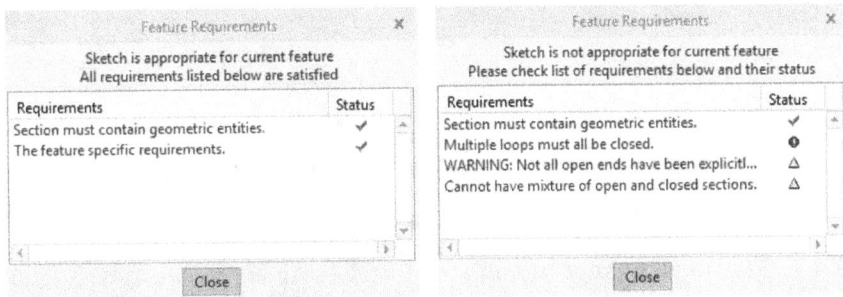

Fig. 4.10.   Feature Requirements — report for a good sketch (left), report for a bad sketch (right).

The **Inspect** group (Figure 4.9) has four diagnostic tools that can check for conformity with the above rules and identify some problems, as follows:

Feature Requirements ( ) tool shows a report of whether a sketch meets the feature requirements (Figure 4.10).

Shade Closed Loops ( ) tool shades the area of a closed-loop sketch in pink, as shown previously in Figure 4.6. This is an indication that the sketch is suitable for solid Extrude, Revolve, Sweep or Blend features.

Overlapping Geometry ( ) tool (Figure 4.9) diagnoses overlapping geometrical entities and/or multiple loops present in a sketch, and highlights entities that overlap.

Highlight Open Ends ( ) tool indicates the loose ends with small red squares, as shown in Figure 4.11.

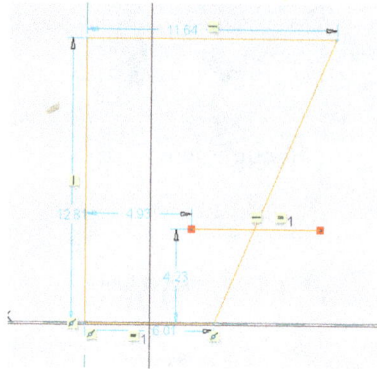

Fig. 4.11.   A sketch with multiple loops indicated with small red squares.

4.4 Creating the Bracket Part — Extrude, Plane and Mirror

4.4.1 *Start a new part model*

(1)   Start Creo (unless it is already running).

(2)   From the Start Up interface, set up a Working Directory, i.e. **File > Select Working Directory** > C:\USER\CREO_PRACTICE.

(3)   Click on **File ≥ New**, or directly click on the **New** ( ) icon from the Quick Access Toolbar to start a new part model. The **New** window will open (Figure 4.12). By default, **Part** (model type) is selected in the **Type** area. Also, a default part name is given in the **Name** box. Delete it and type BRACKET as part name.

(4)   Remove the tick mark of **Use default template** if you are unsure what the default template is and then click the **OK** (OK) icon to accept.

(5)   The **New File Options** window appears. Select a template with an appropriate units system, i.e. **mmns_part_solid** for the Metric System of Units (metre, Newton, second), and click **OK** (OK).

After confirmation, the Part mode window with tools for part modelling opens. Notice the three default datum planes (RIGHT, TOP and FRONT) and coordinate system (PRT_CSYS_DEF) in the Model Tree and Graphics area. The three orthogonal datum planes provide references for the model features.

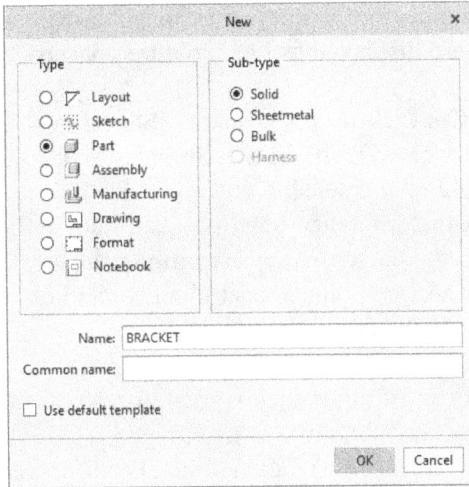

Fig. 4.12.   New part window (BRACKET).

4.4.2 *Creating external sketch and solid extrude*

(6)   Click on the **Sketch** ( ) icon (**Model** tab). The **Sketch** dialogue box opens for the sketch setup. Move the mouse cursor in the area of the three main datum planes. Notice how they flush, indicating preselection, when the cursor hovers over these planes. Click on the **TOP** datum plane to select it as **Sketch Plane** for sketching. Select a **Reference** plane for the sketch orientation (Section 2.7.3). Note that the area active for selection is highlighted in green. Use the **Flip** icon where necessary to change the viewing direction. If both selections are correct, then click on **Sketch** to start.

(7)   The **Sketch** dashboard will appear with all tools (explained in Section 4.3). The **OK** (✔) and **Cancel** (✘) icons are located on the right.

(8)   By default in Creo 4.0, the sketching plane is not aligned to the screen. Click on **Sketch View** ( ) icon (Graphics toolbar) to align.

4.4.2.1 *Using centreline in a sketch*

Centreline promote symmetry when sketching. In this example, a rectangle will be sketched symmetrically against the vertical and horizontal references. Centrelines are displayed as magenta-coloured dashed lines while references are displayed as blue-coloured dashed lines.

(9) Click on **Centreline** (¦) icon (**Sketching** group) to start. Alternatively, click RMB in the Graphics area to open the **Sketch tools** mini menu and select **Centreline**. Position the cursor to snap to the horizontal reference line (a dashed line) and click on the first point. Move the cursor away, snap to the same reference and click again on the second point. A centreline coinciding with the horizontal reference is created.

For simplicity, <u>click on</u> or <u>click</u> is used instead of '<u>click and release</u>'.

(10) Select **Centreline** (¦) again and sketch a vertical centreline following the previous instructions.

4.4.2.2 *Using line chain in a sketch*

(11) Click on the **Line Chain** (⌄) icon (from the **Line** drop-down menu in the **Sketching** group) to start. Move the cursor to the upper-left quarter and click on the first point. Move the cursor down vertically across the horizontal centreline and continue until the **Symmetric** constraint icon appears. Click on the 2nd point. Note the **Vertical** and **Symmetric** constraint icons.

(12) Continue the **Line Chain** by moving the cursor horizontally and across the vertical centreline until another **Symmetric** constraint icon appears. Click on the 3rd point. Move the cursor up and, at a certain point, the equal constraint will appear as shown in Figure 4.13 (left). Click on the 4th point. Move the cursor left until it snaps (coincides) to the start point. Press the MMB to stop the **Line Chain** tool.

(13) If the tool has stopped working or a mistake has been made, then press the MMB to interrupt the process and restart it again. If you add a new line, then make sure that the start point snaps with an existing line point.

Notice how the Sketcher attempts to stick the cursor to the existing entities imposing coincident constraint.

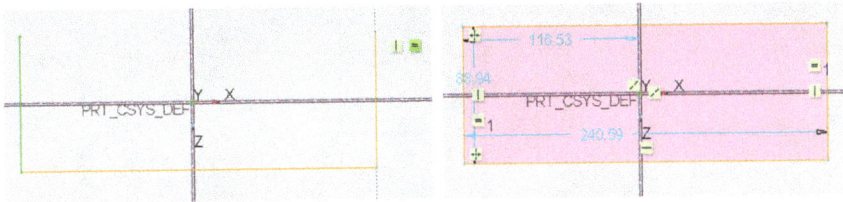

Fig. 4.13.   Sketching the base.

(14) Click the MMB twice to stop sketching. Dimensions will be placed automatically as shown in Figure 4.13 (right). The section is shaded in pink, indicating a section with closed loop that is appropriate for the solid **Extrude** feature. Notice that the two left vertices are symmetric against the horizontal centreline, all lines are horizontal or vertical and the right vertical line is equal to the left one. The Sketcher automatically applies these constraints.

  If the required constraints are not present, then they can be assigned manually using the tools from the **Constrain** group.

(15) For example, to apply symmetric constraints to the left and right vertices against the vertical centreline, click on **Symmetric** icon (**Constrain** group), next click on the vertical centreline and then click on the two corresponding vertices.

Look at the message line at the bottom of the screen to understand what the correct inputs and their order for a particular tool are.

Any constraint can be deleted as follows: click on the **Select** ( ) icon to activate, then select the constraint(s) and press the DELETE key.

During the sketching process, the Sketcher evaluates the geometry, dimensions and constraints. If there is a problem, the **Resolve Sketch** dialogue box will appear, as shown in Figure 4.14, inviting the user to delete unwanted constraint or dimension from the list.

  Notice that after applying a symmetric constraint to the 4 vertices in Figure 4.13, the number of dimensions has reduced from 3 to 2.

(16) Click on the **Select** ( ) icon from the ribbon (**Operations** group), then double-click the vertical dimension until it highlights, type

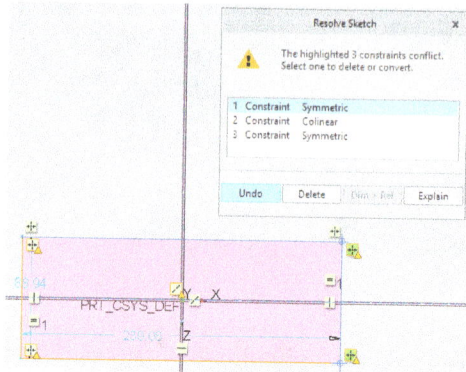

Fig. 4.14.   Resolve Sketch dialogue box.

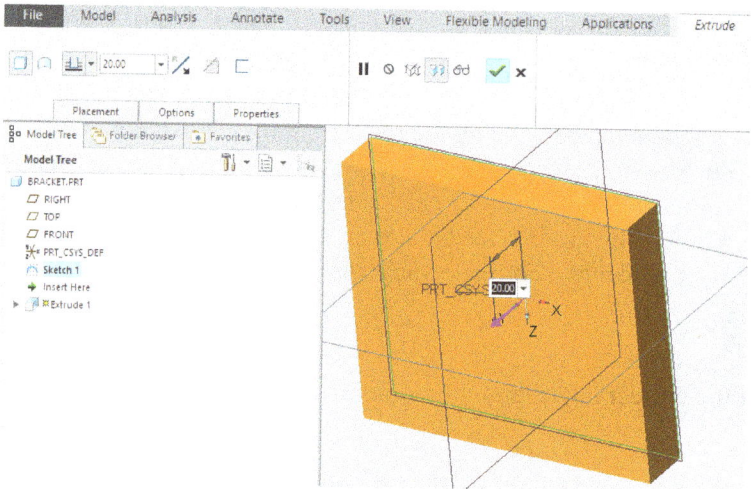

Fig. 4.15.   Extrude feature with an external sketch.

100 mm and press the ENTER key to finish. Double-click on the horizontal dimension until it highlights, type **120** mm and press ENTER. The sketch will automatically update.

(17) Click on **OK** (✔) to close the **Sketch**.

(18) Press (MMB + Hold) and then drag the mouse to spin the part.

(19) Select **Standard Orientation** from the Graphics toolbar to change the orientation to isometric.

(20) With the **Sketch 1** still selected, click on the **Extrude** (⬜) icon. The **Extrude** dashboard opens as shown in Figure 4.15).

(21) Click and drag (LMB + Hold) the depth handle (small white square) to **20** mm or type **20** mm distance in the depth area of the dashboard.

(22) Click on **OK** (✔) to close. **Extrude 1** feature appears in the Model Tree.

4.4.3 *Creating datum planes* (*offset plane*)

Sometimes, the three main planes (RIGHT, TOP and FRONT) are insufficient or inconvenient for the creation of specific geometry. In such cases, additional datum planes can be introduced as supporting references. Creo has a command **Plane**, which creates a datum plane by selecting a combination of geometrical entities that fully define a plane in 3D, such as datum planes, points, axes, edges, vertices, etc.

The next procedure will demonstrate how to create an offset datum plane for sketching the bracket support.

(23) Select the RIGHT datum plane and click on the **Plane** ▱ icon from the mini menu. (The same feature icon is present in the **Datum** group, model ribbon. A new datum plane, parallel to RIGHT, will be created.

(24) Click and drag (LMB + Hold) the offset handle (white grip handle) of the new datum plane to **35** mm offset (Figure 4.16).

(25) Click on **OK** in the **Datum Plane** dialogue box to close.

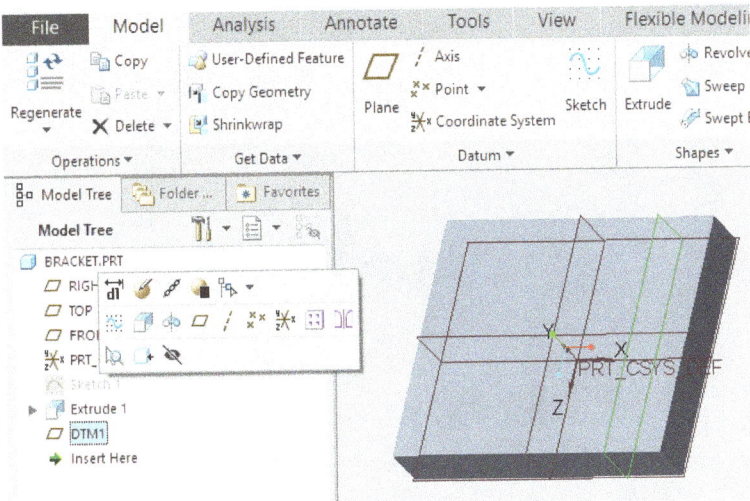

Fig. 4.16.   Creating offset datum plane.

4.4.4 *Creating an internal sketch and extrude feature*

(26) Click on the offset datum plane **DTM 1** either from the Graphics area or from the Model Tree as shown in Figure 4.16.

(27) Click on **Extrude** (⬜) icon either from the mini menu or from the model ribbon. The Sketcher with its ribbon will open.

(28) Click **Sketch View** (⬛) to orient the sketching plane parallel to the screen.

(29) Draw two coaxial circles as shown in Figure 4.17. To do this, click on the **Circle** (◯) icon (**Sketching** group) to activate. (Another way is to press (RMB + Hold) and activate **Circle** tool from the **Sketch tools** mini menu.) Position the cursor to snap at the horizontal reference, click on a point to place the centre, move the mouse outward, and then click on a point to place the radius. Draw a second circle snapping its centre with the centre of the first circle.

(30) Click the MMB to stop the **Circle** tool. Notice that dimensions are created automatically. The section is also shaded in pink, indicating a closed-loop section.

(31) Double-click on each dimension until it highlights, type the corresponding sizes — **15** mm, **35** mm and **100** mm — and press ENTER after each input.

(32) Press (RMB + Hold) to activate **Sketch Tools** mini menu and click on the green tick (✔) icon to save and close the Sketcher. This will

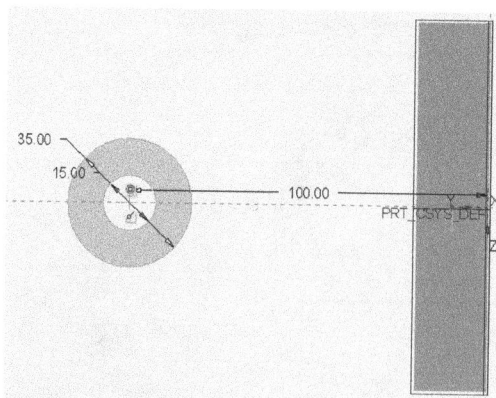

Fig. 4.17.   Sketching a section with two coaxial circles.

bring back the **Extrude** dashboard that is still open. Alternatively, click on the **OK** icon from the ribbon.

(33) Click on **Saved Orientations** ( ) from the Graphics toolbar and select **Standard Orientation**.

(34) Create an extruded protrusion in two directions. Click on the black arrow next to Depth Options ( ) icon, in the **Extrude** dashboard, to open the drop-down menu and select **Both Sides** ( ). Drag the depth handle (white square) to **20** or type **20** mm as the distance.

(35) Click on **Refit** ( ) from the Graphics toolbar.

(36) Click on the green tick ( ) icon, in the **Extrude** dashboard to save and close the feature.

4.4.5 *Creating a complex internal sketch*

(37) Select the offset datum plane **DTM 1** either from the Graphics area or from the Model Tree.

(38) Click on **Extrude** ( ) icon from the mini menu.

(39) Click on the ( ) icon.

 Project ( ) and **Offset** ( ) tools copy (or project) selected items from existing features (edges, points, datums) into the current sketch, thus creating useful references.

(40) Click on the **Project** ( ) icon from the **Sketching** group. Next, click on the outer circle <u>once</u> to project it. A yellow arc will be projected as a result. Click on the other side of the circle to project a second arc. In Creo, each circular edge is interpreted as two arcs. Click <u>once</u> on the top horizontal edge of the base to project it onto the sketching plane. <u>Do not click twice</u> on the same edge because it will create another copy of the same edge. The overlapped geometry will generate an error and will not allow you to exit the Sketcher. To correct, delete the redundant entity or delete all and start again. Figure 4.18 shows the **Type** dialogue box (**Project** tool) and the sketch with three projected entities (two arcs and one line). The default is **Single**, i.e. allowing single one by one selection. **Chain** and **Loop** allow selection of a set of entities that form a chain or a loop.

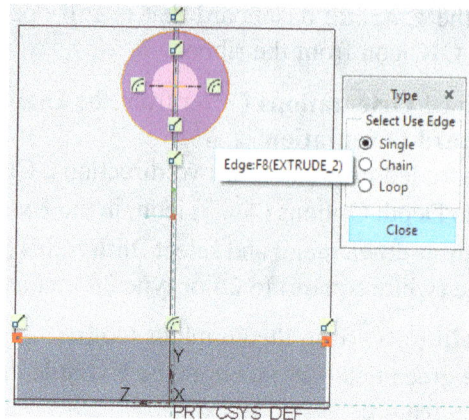

Fig. 4.18.   Use of Projection tool (Type dialogue box).

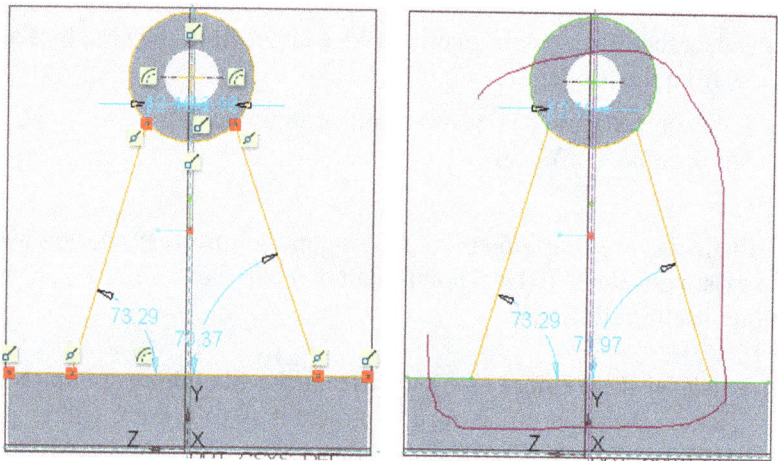

Fig. 4.19.   Bracket support section before (left) and after Delete Segment (right).

(41) Click on **Line Chain** and draw a line on the left-hand side of the centreline, starting from the projected line and up to the left arc as shown in Figure 4.19. Notice how the endpoints snap to the projected entities. Press MMB to stop. Move the cursor to the right and draw another line on the right-hand side. Press MMB to stop **Line Chain**.

(42) Click MMB again to stop sketching. Notice that dimensions are created automatically. The section has multiple loops and six open ends,

indicated with red squares; thus, it cannot be used for an **Extrude** feature unless corrected.

(43) To correct the section, click on the **Delete Segment** ( ) icon, position the cursor in the section, press (LMB + Hold) and then drag the cursor along a trajectory to cross and delete the unwanted entities.

(44) Apply **Symmetric** constraints to the four endpoints, as is explained in the following points.

(45) Click on **Centreline** ( ), then click on two locations on the vertical dash dot line, snapping the point to it, to draw a vertical centreline.

(46) Click on the **Symmetric** icon (**Constraint** group), click on the vertical centreline first and then click on the upper two endpoints. Click on the vertical centreline again and then select the two lower endpoints. All four endpoints are made symmetrical against the vertical centreline. Notice the constrain labels in the sketch (Figure 4.20, right). The number of dimensions have reduced from four (Figure 4.20, left) to two (Figure 4.20, right).

(47) Click on the **Select** ( ) icon, double-click on the angle dimension to activate, type **70** and press ENTER. Double-click on the horizontal dimension, type **80** mm and press ENTER.

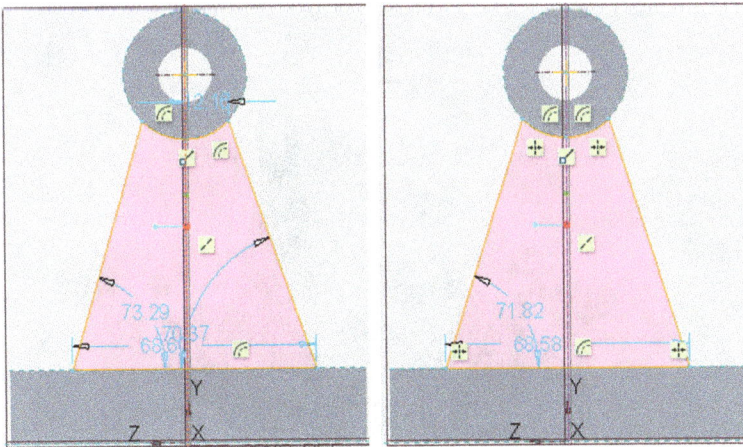

Fig. 4.20.   Bracket support section before (left) and after (right) applying symmetrical constraints.

(48) Press (RMB + Hold) to activate the **Sketch tools** mini menu, and then click on **OK** (✔) to close the **Sketch**. This will close the Sketcher and bring back the **Extrude** dashboard, which is still open.

(49) Click on the **Saved Orientations** ( ) icon from the Graphics toolbar and select **Standard Orientation**.

(50) From the **Extrude** dashboard, click on the Depth Options ( ) drop-down menu and select **Both Sides** ( ). In the Graphics area, click and drag (LMB + Hold) the depth handle to **10** or type **10** mm depth directly in the dashboard.

(51) Click on the **OK** (✔) icon to close the feature.

(52) Click on **Refit** ( ) icon from the Graphics toolbar (Figure 4.21).

(53) Click on **Save** (Save) icon to save the latest model modifications.

4.4.6 *Mirror command — creating the symmetrical support*

(54) Press (CTRL + Hold) and click on **Extrude 2** and **Extrude 3** from the Model Tree to select both. The mini menu appears next to the selected features.

(55) Click on the **Mirror** ( ) icon from the mini menu (Figure 4.22), or from the ribbon (**Editing** group).

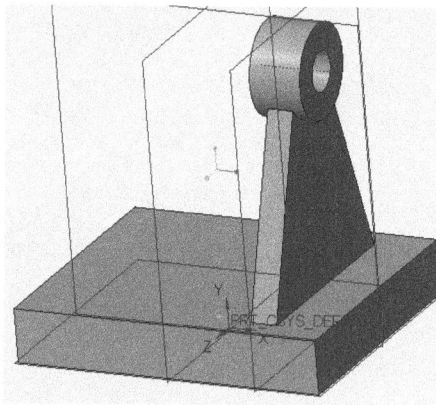

Fig. 4.21.   Completed Extrude feature (bracket support).

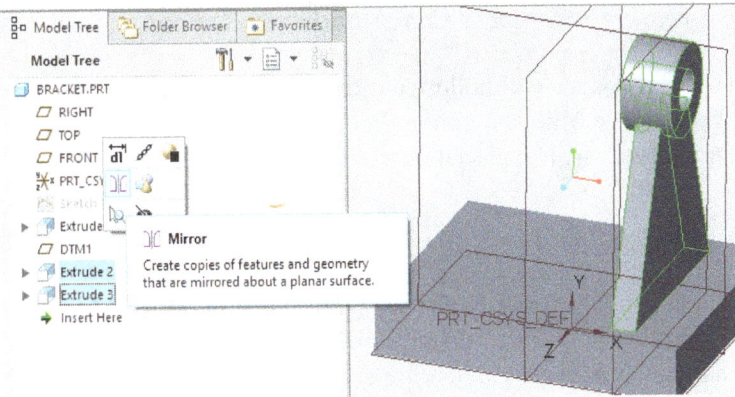

Fig. 4.22. Selection of features, the mini menu, and Mirror command.

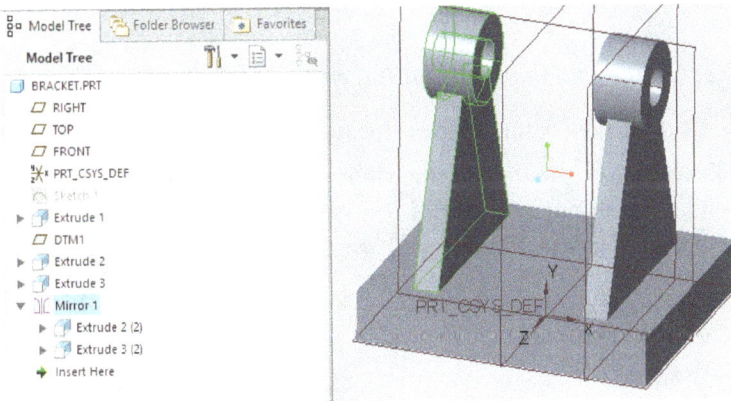

Fig. 4.23. BRACKET after completion of Mirror command.

(56) Select RIGHT datum plane (plane of symmetry). Click on the green tick (✔) in the **Mirror** dashboard to save and close. The model will look as shown in Figure 4.23.

Creo has a number of editing commands that can duplicate features, including Mirror, Copy/Paste, and Pattern. The previous example demonstrates how parts with symmetrical geometry can be modelled more efficiently.

(57) Click on **Save** (Save) from the Main Toolbar to save the model.

4.4.7 *Creating a complex sketch and extruded cut*

(i) The next feature will hollow out the BRACKET base using **Extrude** with the **Remove Material** option. In order to create the desired shape for the cut, its sketch will be placed on a plane offset from the top base surface.

4.4.7.1 *Create another offset datum plane*

(58) Click on the **Plane** icon (▱), **Datum** group, i.e. **Model tab > Plane**. Select the top surface of the bracket base and drag the offset handle downward to **10** mm (or type in directly **10** mm in **Translation** slot, **Datum Plane** window), as shown in Figure 4.24. Click on **OK** to finish. This new datum plane will be used as a sketching plane.

4.4.7.2 *Create extrude with an internal sketch*

(59) Select the new datum plane (unless already selected) and click on the **Extrude** (▱) icon from the ribbon. Notice that the Sketcher opens and this datum plane (**DTM 3**) is selected as the sketching plane.

(60) Click on **Datum Planes** (▱) from the Graphics toolbar to disable their display. Click on **Sketch View** (▱) and **Hidden Line** (▱) from the Graphics toolbar.

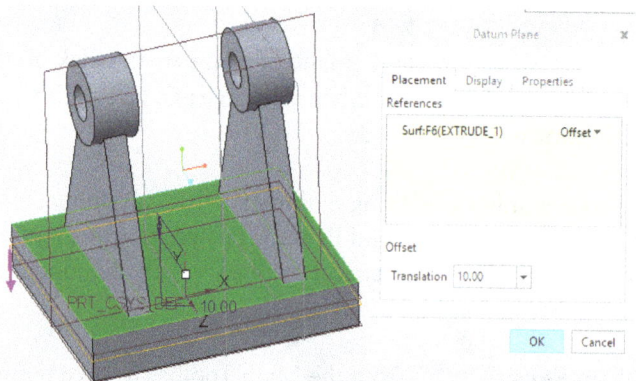

Fig. 4.24.   Offset datum plane.

(61) Sketch one vertical and one horizontal centrelines to snap with the two main datum references as follows: click on the **Centreline** ($\vdots$) icon, **Sketching** group (do not confuse it with the geometry centre-line in **Datum** group) and click on two points on the vertical refer-ence. Notice how these points snap to the reference line. Sketch another centreline to snap on the horizontal reference line.

(62) Click on the **Rectangle** ($\square$) icon, **Sketching** group. Draw a rectan-gle symmetrical against the horizontal and vertical centrelines. Click on a point in the upper left quadrant to start sketching the rectangle upper left corner. Slowly move the cursor horizontally to the right until the symmetry (constraint) icon appears. Now move the cursor vertically down until another symmetry icon appears and click on a point to define the rectangle's lower right corner. Try to avoid snap-ping the endpoints to any hidden geometry entity (Figure 4.25).

Always <u>click</u> or <u>click on</u> a point ('<u>click and release</u>') technique when sketching geometry as opposed to holding down the left mouse button (LMB + Hold) and dragging the mouse to modify an entity.

(63) Click MMB to stop sketching and to activate the **Select** tool ( ).

(64) Double-click on the horizontal dimension to edit, type **110** mm and press ENTER. Double-click on the vertical dimension, type **90** mm and press ENTER.

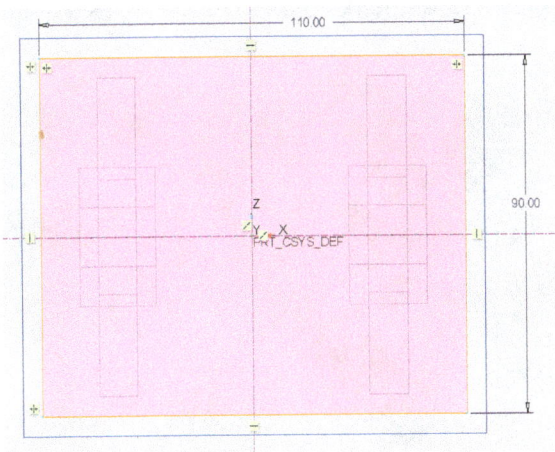

Fig. 4.25. Rectangle symmetrical to the main references (notice the symmetry con-straints and dimensions).

4.4.7.3 *Sketching the arcs*

The **Arc** tool can create **3-Point**, **Tangent End/Center and Ends** arcs. In this exercise, a **Center and Ends** arc will be created.

(65) Continue the previous sketch (rectangle) by sketching an arc at the rectangle corner. Click on the black arrow next to the **Arc** ( ) icon to open the drop-down menu and click on **Center and Ends** option. Click on the rectangle's upper left corner (vertex) to define the arc centre. Make sure that it snaps to the vertex. Move the cursor downward to define the radius, then click on a point that snaps to the vertical left side. Move the cursor anti-clockwise and click on a second point that snaps to upper horizontal line to complete the arc. Click MMB to stop the Arc tool (Figure 4.26). Double-click on the radius dimension to activate, then type **16** mm and press ENTER key.

4.4.7.4 *Mirroring the arcs in the sketch*

(66) Continue the sketch by mirroring the arc. Click LMB on the arc to select it. If selected, it will appear in green colour. Click on the **Mirror** icon, **Editing** group, and then click on the vertical centreline. The arc will be mirrored.

(67) Press (CRTL + Hold), click on the two arcs to select them both and repeat the **Mirror** procedure using the horizontal centreline. Look at the message line, at the bottom of the screen, as a reminder of what

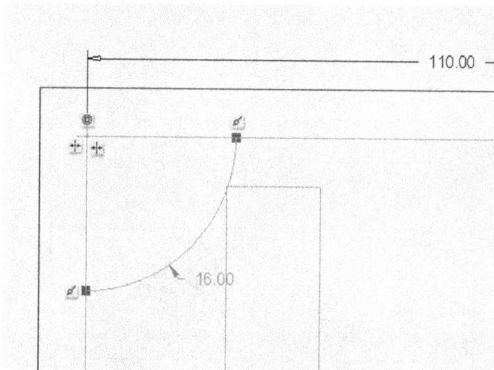

Fig. 4.26.   Center and Ends arc.

should be the next input. The **Mirror** tool is available only when there is a centreline and selected entity(ies) in the sketch. Use **Undo** or start again if you make a mistake.

The current sketch has multiple loops and cannot be used in Extrude. To correct it, the corners should be trimmed.

(68) LMB click on the **Delete Segment** icon ( ), **Editing** group. Position the cursor on the left corner, then press (LMB + Hold) and drag the cursor across the lines that lie on the outer side of the arcs to trim. Perform the same function on all four corners.

(69) Click MMB to stop the tool. If the section is closed, then it will be filled in pink colour (Figure 4.27). If this is not the case, then either the sketch is opened or there are multiple loops present. Small red squares indicate the loose ends of these areas.

To tackle contour problems, use the **Inspection** tools, the **Inspect** group: **Shade Closed Contours** ( ) tool to identify loops, highlight **Open Ends** ( ) and **Overlapping geometry** ( ) tool for intersecting lines. These tools can be toggled ON/OFF as required. The first two are ON as default. If a sketch problem cannot be identified, then the easiest way to correct is to delete all drawn entities and start again. (If the user has deleted the horizontal and vertical main references, then these can be selected back using the **References** (References) tool, **Setup** group.

4.4.7.5 *Creating sketch dimensions*

Many different dimension types can be created with the **Dimension** tool. To create a dimension, start the **Dimension** tool, select the item(s) to be dimensioned and then press the MMB to locate the dimension value.

(70) Notice that after selecting the **Delete Segment** and MMB click (in the previous section), the dimensions change their location and need to be redefined.

(71) Click LMB on the **Dimension** icon ( ) to start. Select the top and bottom horizontal lines with LMB click, position the cursor between the lines and click MMB to place the dimension. The **Resolve**

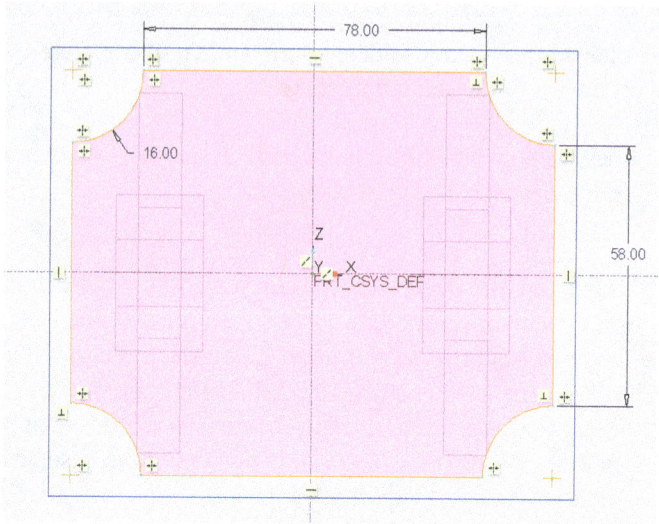

Fig. 4.27.   Pink colour showing that sketch is closed.

Sketch dialogue box opens to indicate that there are redundant dimensions. Keep the **90** mm vertical dimension and delete the 58 mm to resolve the conflict.

(72) With the **Dimension** (⊢→⊣) tool still active, select the left and right vertical lines, position the cursor between the lines and click MMB to place the dimension. Resolve the sketch conflict by keeping the **110** mm dimension and deleting the current one.

(73) Click on the **OK** (✔) icon to close the Sketcher.

(74) There will be a message that the **Extrude** feature will switch to **Remove Material**.

4.4.7.6 *Extrude* (*with remove material*) *and depth options*

ⓘ The **Remove Material** icon (▱) in the **Extrude** dashboard can be toggled ON/OFF as required to **Remove Material** (see Figure 4.28). The default option is to add material. If the sketching plane is inside a solid feature, then the command will automatically switch to **Remove Material**, as in this case. A magenta arrow indicates the direction for material removal — inside or outside the section. To change direction, click on the magenta arrow or click on the direction icon in the dashboard.

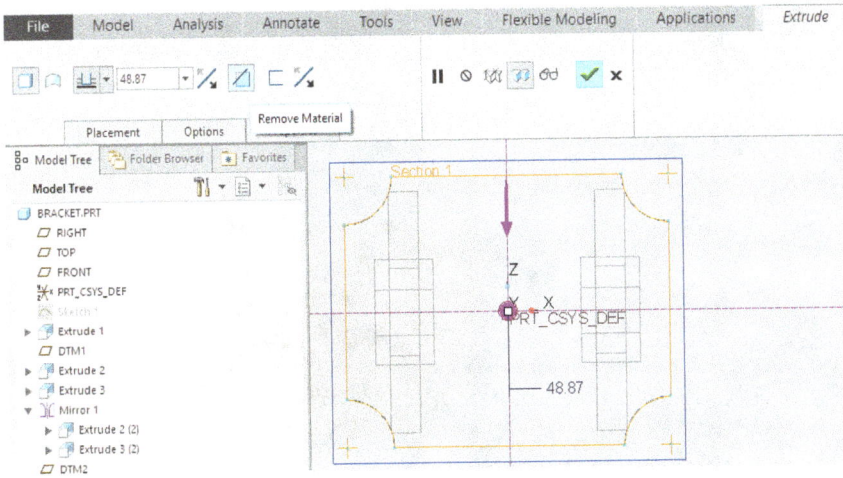

Fig. 4.28.   Extrude command dashboard (Remove Material).

(75) Click on **Shading** or **Shading with Edges** ( ), in the Graphics toolbar.

(76) In the **Extrude** dashboard, click on the **Remove Material** icon ( ) to toggle to ON (unless already switched ON automatically).

(77) Click on the black arrow next to the depth options ( ) icon and from the drop-down menu (Figure 4.29) select **Extrude to Intersect with All Surfaces** ( ) icon. The active extrude directions are indicated by magenta arrows. Click on all arrows and direction icons in the dashboard to view the result and learn their function.

The default depth option is one-sided extrude ( ) with a fixed depth value. This option will work if the depth value is larger than the thickness to cut through. However, the **Extrude to Intersect with All Surfaces** ( ) option is a more efficient choice that will work with varied thickness to cut.

(78) Click on **Refit** ( ), Graphics toolbar, to view the final result.

(79) Click on the green tick ( ) in the dashboard to complete the feature.

(80) Click **Save** ( ) from the toolbar to save the model.

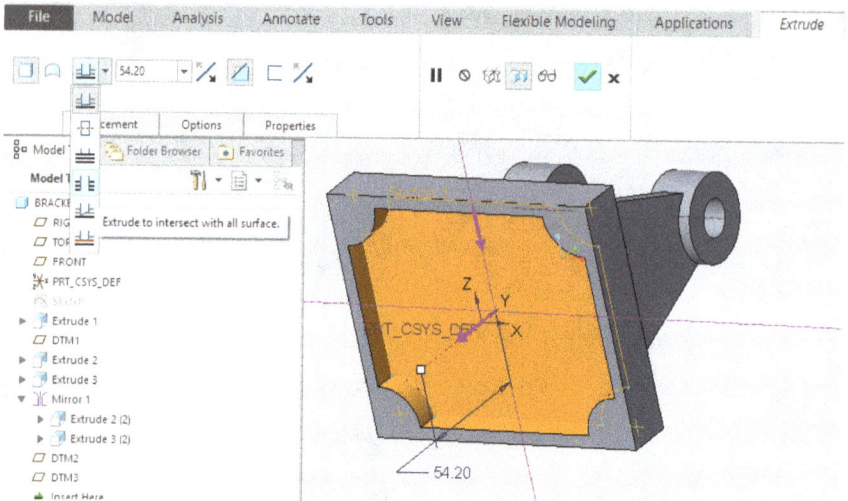

Fig. 4.29.   Extrude with Remove Material (Through All depth option).

4.4.8 *Creating rounds and chamfers with multiple references*

Press (CTRL + Hold) and click (LMB) on multiple edges for a single **Round** or **Chamfer** set.

4.4.8.1 *Create rounds on the four edges*

(81) Click on the **Round** (Round ▼) icon from the ribbon.

(82) Press (CTRL + Hold) and click on the four external edges of the base, as shown in Figure 4.30.

(83) Click (LMB + Hold) on the radius handle (the white square) and drag to **10** mm or double-click on the radius value and type in **10** mm.

(84) Click on the green tick (✓) to complete the feature.

(85) Click again on the **Round** (Round ▼) icon. Press (CTRL + Hold) and with the LMB select the four internal edges (Figure 4.31). Click on the radius value slot in the dashboard and enter **2** mm (radius).

(86) Click on the green tick (✓) to complete the feature.

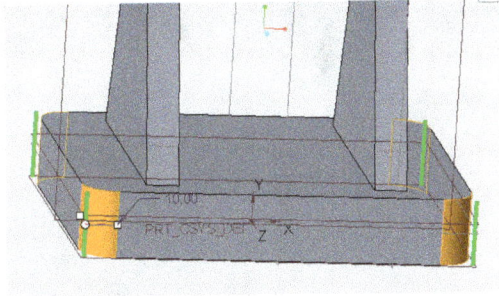

Fig. 4.30.   Round applied to the four edges of the base.

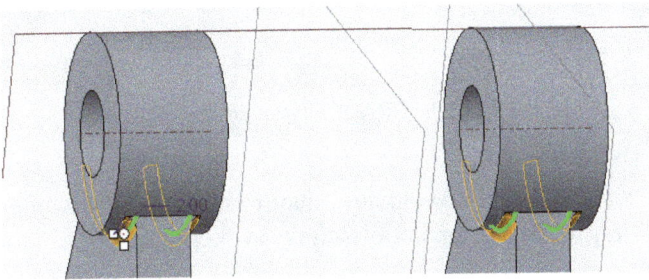

Fig. 4.31.   Round applied to the four internal edges (BRACKET).

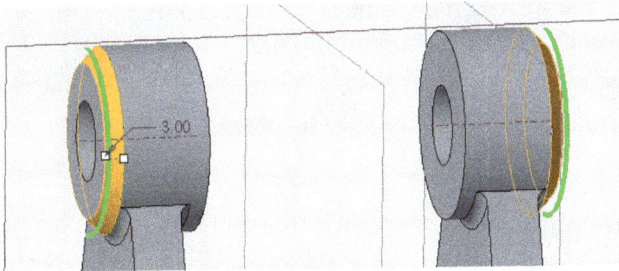

Fig. 4.32.   Chamfer applied simultaneously to the two external support edges.

4.4.8.2 *Creating chamfers*

(87) Click on the **Chamfer** (Chamfer ▼) icon from the ribbon to activate.

(88) Press (CTRL + Hold) and with the LMB select the two external edges, as shown in Figure 4.32.

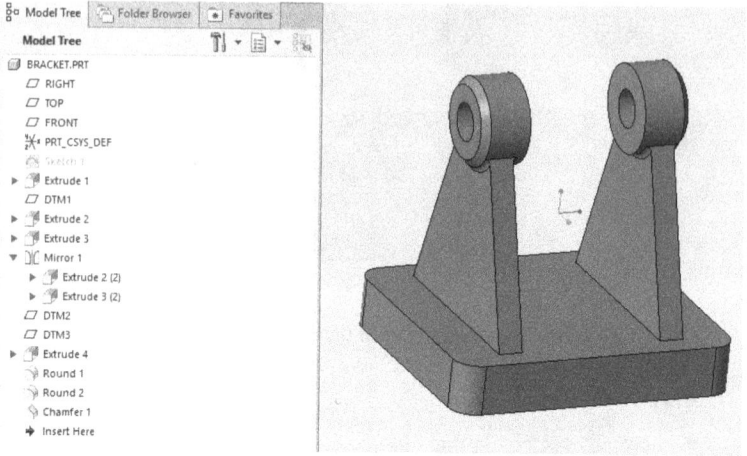

Fig. 4.33.   The BRACKET part and Model Tree.

(89) Click (LMB + Hold) on the chamfer handle (the white square) and drag to **3** mm or double-click on the chamfer value and type **3** mm.

(90) Click on the tick (✔) icon to complete the feature (Figure 4.33).

Note that if the chamfer value is more than **3** mm, it will overlap with the **2** mm round and the chamfer will fail.

(91) Click on **Save** (💾 ^Save^) to save the model.

(92) Click on **File > Close** (✕ ^Close^) and then select **Erased Not Displayed** to remove all models from the current session (RAM).

Round and **Chamfer** are features that are applied directly to the existing geometry. In product design, they are often used for cosmetic reasons to improve the appearance. In mechanical design, these features could serve certain functional or technological requirements following the 'design for manufacture' rules. For example, they can improve the structural strength and facilitate casting and moulding.

Having too many rounds and chamfers could slow down the model regeneration process. Also, adding these features too early in the Model Tree can lead to the following:

- Generate too many edges and vertices that could be wrongly selected as references for subsequent features.
- Create undesirable Parent–Child links and confuse the user.

The recommendation is to create the rounds and chamfers towards the end of the modelling process unless there is a good reason to use these earlier as main shapes.

The development of the BRACKET continues in the next chapters.

4.5 Exercises

Exercise 1
Create a solid part as shown in Figure 4.34 with uniform wall thickness of 5 mm. Use extrude features to add or remove material. Notice the part symmetry. In the sketches, draw construction centrelines and use **Symmetric** constraints. In addition, apply the **Project** tool to copy existing edges as references and create a resilient-to-modification model.

Exercise 2
Create the part shown in Figure 4.35. Use two-sided symmetric and asymmetric extrudes for the two cylindrical shapes and symmetric **Extrude** for the **6** and **8** mm ribs.

Exercise 3
Create the part shown in Figure 4.36. Notice the part symmetry. Use **Extrude** feature to create the base, symmetrical to a vertical and

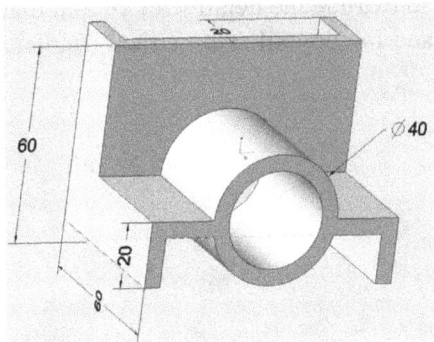

Fig. 4.34.   Solid part — Exercise 1.

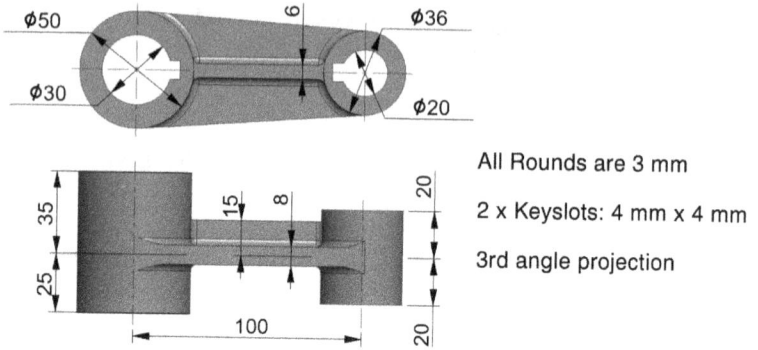

All Rounds are 3 mm

2 x Keyslots: 4 mm x 4 mm

3rd angle projection

Fig. 4.35.   Drawing of the part in Exercise 2.

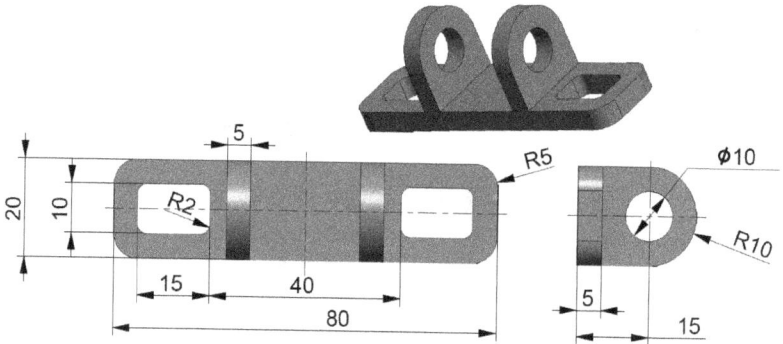

Fig. 4.36.   Drawing of the part in Exercise 3.

horizontal reference. Create an offset datum plane (from the base centre) and use that plane to extrude the detail with **10** mm diameter hole. Apply the **Mirror** command to copy all symmetrical details. All rounds should be done at the end.

Chapter 5

Creating Solid Part Models Using Basic Commands II

5.1 Introduction

This lesson will help the reader to learn how to create parts with more complex geometry. New features and modelling techniques will be introduced to build upon the knowledge from the previous chapters. The modelling process will be illustrated by means of practical exercises to create new and continue to develop parts from the previous chapters, in a step-by-step instructions format.

Aim:

- To continue developing part modelling skills with Creo.

Outcomes:
At the end of this, lesson the reader should be able to:

- Draw more advanced 2D sketches and create a **Revolve** feature;
- Create datum **Plane** and datum **Axis**;
- Create a variety of holes using the **Hole** feature;
- Create ribs using **Profile Rib** feature;
- Duplicate features using **Pattern** and **Mirror** commands.

5.2 Creating Pulley Part — Revolve Command

Figure 4.1 shows the final appearance of the Pulley part. The main shape will be developed using a **Revolve** feature.

5.2.1 *Revolve feature*

ⓘ The **Revolve** command (✤ Revolve) creates a solid feature by revolving a sketched section (a sketch) around an axis of revolution. This axis can be a datum centreline created in the sketched section (internal axis of revolution) or an external datum axis such as straight edge, curve and coordinate system axis (external axis). The user can either select a sketch first and then start the **Revolve** command or start the **Revolve** and then create (or select) a sketch within the **Revolve** dashboard (Figure 5.5).

The sketch must be a closed section for solid features; otherwise, the revolve feature will result in a revolved surface.

Similar to **Extrude**, the **Revolve** dashboard has options to either **Add** (default) or to **Remove Material**. Click on the **Remove Material** icon (◩) to switch to remove material.

The **Thicken Sketch** (⊏) icon is available to create a thin solid revolve from either an open or closed section. (The **Thicken** option must be active before sketching the section as an opened contour.)

🖊 There are some important rules to keep in mind when using **Revolve**: (a) the revolution axis should be a <u>datum</u> **Centreline** (the **Datum** group) and <u>not a construction</u> centreline; (b) the sketched section should not cross the revolution axis; (c) the centreline should lie on the same sketching plane.

Follow the next steps and create the pulley part.

(1) Start Creo. Set a Working Directory, i.e. **File > Select Working Directory** > C:\USER\CREO_PRACTICE.

(2) Start a new part: Click on the **New** (⬜) icon from the Quick Access Toolbar, or **File > New**. In the pop-up **New** window, keep **Part** (<u>Part mode</u>) and **Solid** selected as **Type** and **Sub-type**. Type the name PULLEY. Clear the **Use default template** box and then click on the **OK** (OK) icon to accept.

(3) Select part template from **New File Options** window, i.e. **mmns_ part_solid** for the Metric System of Units (metre, Newton, second), and click on (OK) to accept and close.

(4) The **Model** ribbon opens and the Graphics area displays the three main datum planes (**RIGHT, TOP** and **FRONT**) and the coordinate system (**PRT_CSYS_DEF**). These four default features are brought by the part template.

5.2.1.1 *Creating an external sketch of the revolved section*

It is a good idea to create an external sketch for the **Revolve** feature. This will enable the section in **Revolve** feature to be swapped with another section (external sketch) and create an alternative revolved shape.

(5) Select **FRONT** datum plane from the Model Tree.

(6) Click on the **Sketch** icon ( ), **Model** ribbon. Because **FRONT** has been pre-selected (as **Sketch Plane**), the **Sketch** dialogue box will be skipped, the **Sketch** ribbon with all commands will appear and the model will assume default orientation.

If the initial sketch orientation needs to be changed without leaving the Sketcher, click on the **Sketch Setup** (Sketch Setup) icon and select a new **Sketch** plane and/or **Reference** plane. Click on the **Flip** icon to change the viewing direction.

(7) Select **Sketch View** ( ) (Graphics toolbar) to align the sketch plane to the screen (Figure 5.1).

(8) Draw a horizontal <u>datum</u> centreline that coincides with the horizontal reference line and create the axis for **Revolve** feature. To do this, select the **Centreline** icon ( ), **Datum** group (Figure 5.1), and click on two points that snap to the horizontal reference.

(9) Draw a vertical construction centreline. Click on the **Centreline** ( ) icon (**Sketching** group) and then click on two points that snap to the vertical reference line.

(10) Draw an arc. Click on the **Center and Ends** (Center and Ends) icon to start the tool. First, click on a point on the vertical centreline well below the horizontal reference, and then select two end-points

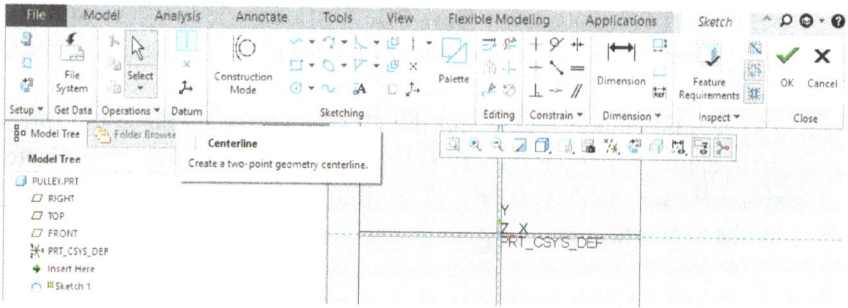

Fig. 5.1.   Datum Centreline (Sketch ribbon).

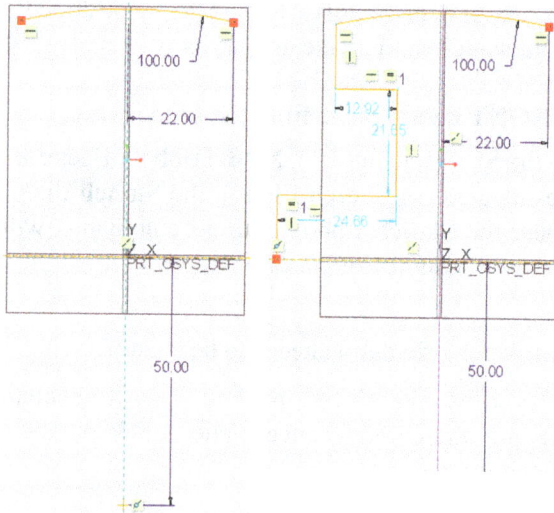

Fig. 5.2.   Sketching the section — an arc (left) and line chain (right).

(Figure 5.2, left). When selecting the arc endpoints, try to make them symmetrical against the vertical centreline. MMB click to finish. The sketch will automatically be dimensioned.

(11) Double-click on the arc radius dimension and enter **100** mm. Double-click on the other dimension and enter **50** mm (distance from the arc centre to the horizontal centreline) and **22** mm (distance from the arc end to the vertical centreline) as shown in Figure 5.2, left.

(12) Click on the **Line Chain** (⌒ Line Chain) icon and draw a series of vertical and horizontal lines. Snap the first point to the left arc end

and the last point to the horizontal centreline as shown in Figure 5.2, right. MMB click to stop the tool. The Sketcher automatically creates all dimensions that define (solve) the shape. Notice the horizontal (⚊), vertical (⊥) and coincident (✍) constraints.

(13) Mirror all lines to create the other side. Press CTRL key and hold it down (CTRL + Hold) and then click on all five lines. All selected lines will now appear in green colour. Click on the **Mirror** (🕮) icon (**Editing** group) and then select the vertical centreline to finish. The sketch should look as shown in Figure 5.3.

(14) To close the section, start **Line Chain** tool and draw a line connecting the open ends (indicated with red squares). MMB to stop the tool. The section should be filled in pink colour, indicating a closed loop. Notice the symmetrical constraint (✛) after **Mirror**.

(15) Accidentally, the Sketcher might create unwanted constraints. To delete, select the constraint icon and press DELETE key. For instance, delete the unwanted equal (═) constraint.

(16) Create dimensions shown in Figure 5.4. Start the **Dimension** (↦) tool and select the entities to dimension. To place a distance between two parallel lines (or centrelines), click on the first line, next click on the second line and finally click the (MMB) between the two lines to place the dimension.

(17) Double-click on each dimension to edit, type the correct value and press ENTER. Type **15** mm for the hub thickness, **50** mm for hub

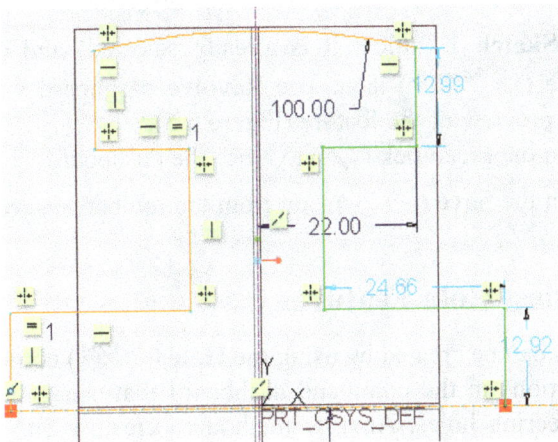

Fig. 5.3. Sketching the section — mirror of all lines against the vertical centreline.

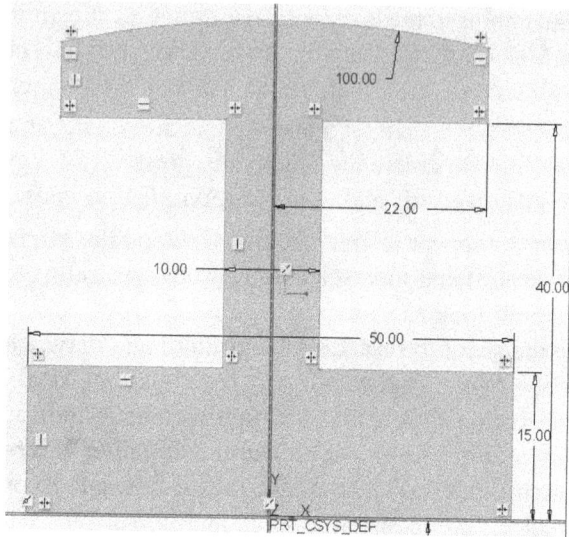

Fig. 5.4.   Sketching the section — dimensioning.

width, **40** mm for distance from the inner side to the centreline and **10** mm for thickness of the middle area, as shown in Figure 5.4.

(18) Press the (RMC + Hold) to activate the **Sketch Tools** mini menu and click on the green tick (✔) to save the sketch and exit.

5.2.1.2 *Revolve feature with an external sketch*

(19) Select **Sketch 1** (unless it is already selected) and click on the **Revolve** (⚙ Revolve) icon. The **Revolve** dashboard will open and show a preview of the feature (Figure 5.5).

(20) Click on the green tick (✔) to close the dashboard.

(21) Click on the **Save** (▥ Save) icon from the toolbar to save the model.

5.3 Creating Hole Features

Hole features can be created by using the **Hole** (▥ Hole) command. There are many options in the command dashboard that can generate a variety of engineering holes. Another method to create a simple cylindrical hole is to use either **Extrude** (sketching a circle) or **Revolve**

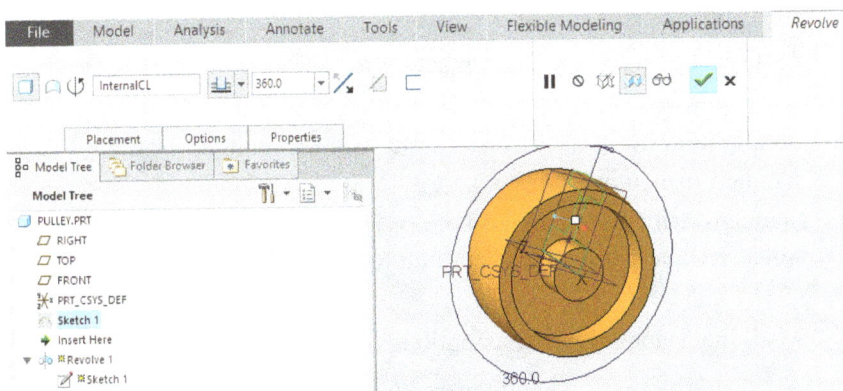

Fig. 5.5.   Revolve dashboard.

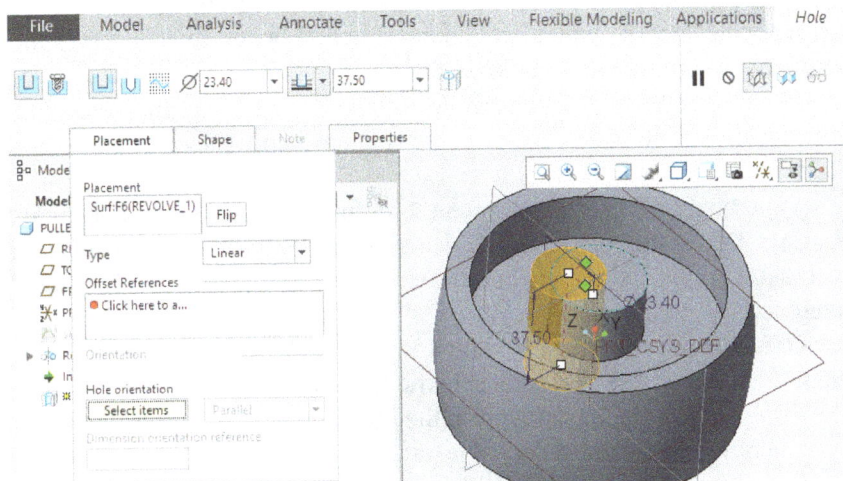

Fig. 5.6.   Hole dashboard and Placement tab.

(sketching a rectangle) features with **Remove material** option. However, direct use of the **Hole** feature is a much more productive way to create engineering holes because it does not require a sketch, it has less dimensioning and it can make variety of holes, including threaded holes, for mechanical design.

To start the command, click on the **Hole** (Hole) icon from the **Model** ribbon, **Engineering** group, i.e. click on the **Model** > **Engineering** > **Hole**. The **Hole** dashboard will appear as shown in Figure 5.6.

Hole types

The **Hole** dashboard has icons that activate the following main hole types:

- Simple hole (![icon]) — A default hole with rectangular profile (![icon]) or standard drilled profile (![icon]).
- Standard drilled and tapped hole (![icon]) — A hole with thread dimensions selected from a list according to ISO or UNC standard series.

Also, the Standard hole has options for counter bore (![icon]), countersunk (![icon]), threaded (![icon]) and tapered (![icon]).

Placement tab

Click on the **Placement** tab (Figure 5.6) to open the pull-down panel in the **Hole** dashboard. This panel contains active boxes for collecting the main hole references and inputs as follows:

Placement box — Click inside this box to activate and select the hole location reference. This can be a plane, an axis or a point (for radial hole).

Type box — A pull-down menu to define the hole type, i.e. **Linear**, **Radial**, **Diameter** or **Coaxial**. It is a context-sensitive menu that will show only the types corresponding to the selected **Placement** reference.

Offset References box — Click inside the box to activate and then input the offset references for a specific hole type.

Flip — Click on the icon to flip the hole direction against the selected placement reference (surface, plane, etc.).

Depending on the selected **Placement** and **Offset References** from the model geometry, the **Hole** command can create the following hole types:

- **Linear** — The hole is positioned on a flat surface, and the hole axis is offset from two other references. Select (a surface, a plane or a point) as reference in the **Placement** box. Next, click inside **Offset References** box to activate, then press (CTRL + Hold) and with the LMB click on two other references, i.e. datum planes, surfaces or edges, to define the offsets. Another way to assign **Offset references**, represented by two green small handles (Figure 5.6) is to pick a green handle with the (LMB + Hold), then drag it slowly until it snaps to an appropriate offset reference and release the LMB. Both offset handles need to be attached to appropriate references.

- **Radial** — The hole is placed on a surface and the hole axis is controlled by a distance from an axis (radius) and an angle from a datum plane. Select the placement reference first (**Placement** box). Next, click inside **Offset References** box to activate, then press (CTRL + Hold) and with the LMB select an axis and a plane (angular reference).

- **Diameter** — This hole is similar to **Radial,** but instead of radius it is controlled by a diameter. Click on a flat surface to select the placement reference (**Placement** box). Then click inside **Offset References** box to activate, press (CTRL + Hold) and with the LMB select an axis and a plane (angular reference) as offset references.

- **Coaxial** — This option creates a hole coaxial to an axis. Click inside the **Placement** box to activate, then press (CTRL + Hold) and with the LMB select the hole placement (surface) and a reference axis.

Note that you need to click inside the **Reference** or **Offset Reference** boxes to activate before selecting corresponding references. Also, press (CTRL + Hold) for multiple selection with the LMB.

Hole diameter
The dashboard has a box to directly enter the hole diameter value for simple holes. For standard drilled and tapped holes, a box with a pull-down list of ISO or UNC series is available with their standard sizes.

Hole depth
Similar to the **Extrude** command, the hole depth options are as follows: drill to specified depth (⯊), up to next (⯊), drill through all (⯊ ⯊), up to selected (⯊) (point, plane, surface…) and drill to intersect with selected (⯊).

The next tasks will illustrate how to use the **Hole** feature in the PULLEY model development.

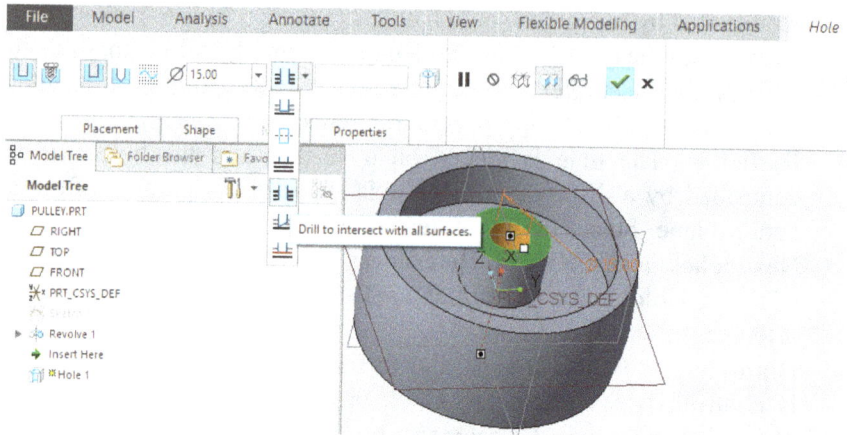

Fig. 5.7.   Hole dashboard — Coaxial hole.

(22) Create a <u>coaxial hole</u> in the pulley hub. Click on the **Hole** ( ) icon from the **Model** ribbon (**Model > Engineering > Hole**) to activate.

(23) Click on **Placement** tab and then **Placement** box to initiate reference collection. Press (CTRL + Hold) and click on the side pulley surface first and then on the pulley axis, as shown in Figure 5.7.

(24) Change the hole diameter to **15** mm and depth to the **Drill to intersect with all surfaces** option. The preview should show the coaxial hole. If the attempt is unsuccessful, click on **Cancel** and start again.

(25) Click on the green tick ( ) icon to close the **Hole** feature.

(26) Click on the **Hole** icon again to create a <u>diameter hole</u>.

(27) Click on the **Placement** tab and then inside **Placement** box to initiate reference collection. Click on the PULLEY middle side surface as placement. In **Type,** click on the black arrow and switch from **Linear** to **Diameter** to create a diameter hole (see Figure 5.8).

(28) Click inside **Offset References** box to activate, then press (CTRL + Hold) and then click on the pulley axis and **TOP** datum plane to select as offset references.

(29) Change the hole diameter to **20** mm, location diameter to **55** mm, angle to **45** degrees and the depth to **Drill to intersect with all surfaces,** as shown in Figure 5.8.

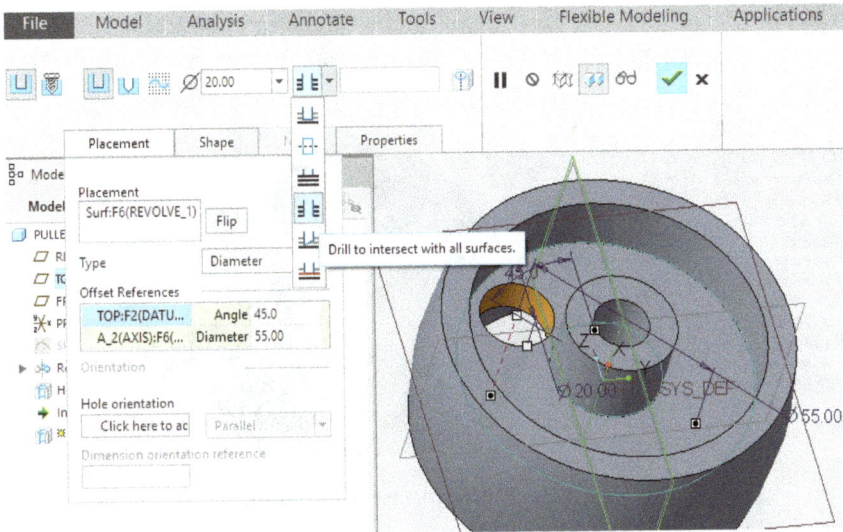

Fig. 5.8. Hole dashboard — Diameter hole.

(30) Click on the green tick (✔) icon to close the **Hole** dashboard.

(31) Click on **Save** (Save) from the toolbar to save the model.

5.4 Creating a Pattern Feature

Pattern features can be used to create multiple copies of a feature (or a **Local Group** of features) as an incremental array.

To create a pattern, <u>first, pre-select</u> the feature to be patterned. For example, click on the **Hole 2** to select, and then click on the **Pattern** ( ) icon in the **Model** ribbon, i.e. (**Model > Editing > Pattern**).

The **Pattern** dashboard opens and the feature parameters are displayed as dimensions (Figure 5.9).

The drop-down menu **Dimension** contains all possible pattern types: **Dimension, Direction, Axis, Fill, Table, Reference, Curve** and **Point**.

The most common pattern types are **Dimension** and **Axis**:

• **Dimension** pattern — This pattern copies a pre-selected feature in one or two directions following the direction of a selected dimension.

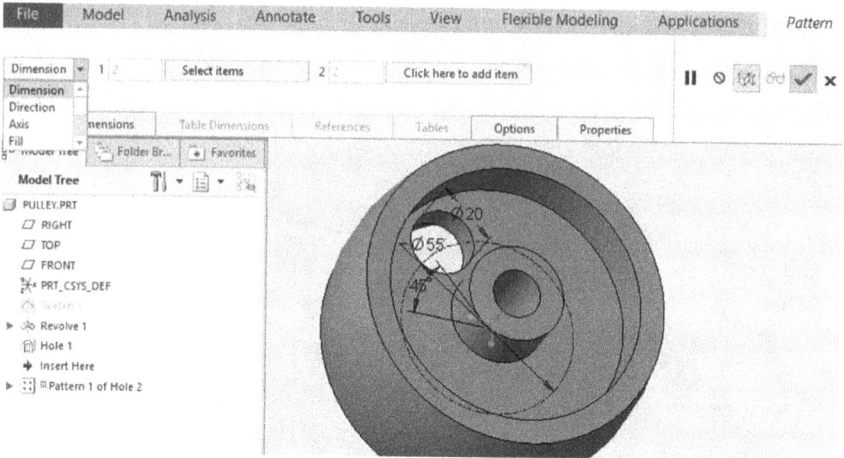

Fig. 5.9. Pattern command dashboard.

By default, **Dimension** type is selected. Click on a linear or angular dimension to add appropriate dimension in **Select items** dialogue box for the first direction and then type in the increment value. Also, type the number of copies for dimension **1** in the dashboard dialogue box (the default is 2). To add a second dimension, click on **Click here to add item** dialogue box to activate the direction **2**. Next, click on a second dimension, type the increment value and specify the number of instances;

• **Axis** pattern — This pattern creates rotational copies. Select **Axis** option from the **Dimension** pull-down menu. Next, select an axis with the LMB click. It could be either a feature axis or a datum axis. Specify the number of instances and the increment value (the angle between pattern members) or the total angle value.

All pattern members, including the initial selected (Parent) feature, are grouped together and act as a single feature that can be mirrored or patterned.

To delete a pattern feature and preserve the Parent feature, use the special command **Delete Pattern** as follows: Select the pattern, click the RMB to open the mini menu, and then click on **Delete Pattern**. If you use Delete command, it will delete all features grouped in the pattern.

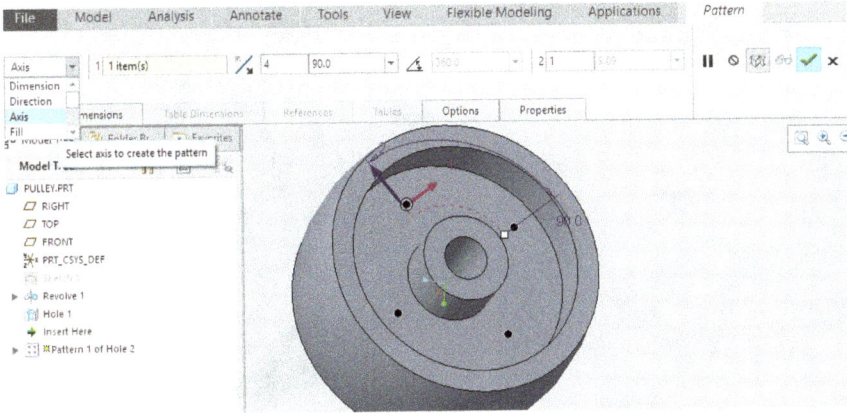

Fig. 5.10. Pattern command with Axis option.

5.4.1 *Creating a rotational pattern (axis option)*

(32) Select **Hole 2** feature and click on the **Pattern** (⊞) icon in the **Model** ribbon, **Editing** group.

(33) Select the **Axis** option from the **Dimension** pull-down menu as shown in Figure 5.10.

(34) Click on the axis of the **Revolve 1** feature. By default, the pattern assumes 4 instances (Figure 5.10).

(35) Enter **6** (the number of pattern members) and **60** (60 degrees angle between pattern members) in the **Pattern** dashboard.

(36) If the pattern is successful, then click on the green tick (✔) to close the feature, otherwise click on **Cancel** and start again.

(37) Finalise the pulley model creating a couple of chamfers. Click on the **Chamfer** (⌐ Chamfer ▾) icon to start an **Edge Chamfer**. Select all four peripheral external edges. Keep the **DxD** type chamfer and type in **2** mm chamfer size.

(38) Create another chamfer feature with the four hub edges with **1** mm chamfer size.

(39) Click on **Save** (⊞ Save) to save the model. Notice the list of features in the Model Tree.

(40) The PULLEY part will look as shown in Figure 5.11.

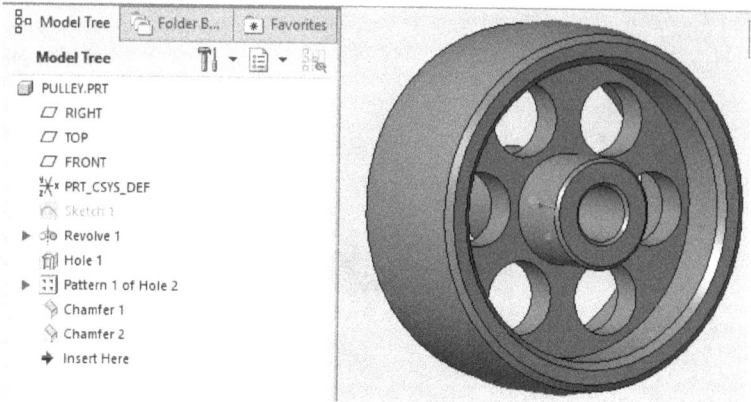

Fig. 5.11.   Pulley part.

5.5 Creating Datum Planes and Datum Axes

Datum features (datums) are basic features that are needed as references when creating sketches, extrudes, revolves, holes and other features for part modelling. Also, datums are required in other downstream Creo applications such as assemblies, mould design, NC (Numerical Control) machining, drawings and others.

Every model (part, assembly, etc.) has three default datum planes and an orthogonal coordinate system at the top of the Model Tree. Often, these main datums are insufficient to build a complex geometry and the CAD designer has to create additional datums.

The commands that are used to create datum features are located in the **Datum** group, **Model** ribbon. They are: **Plane, Axis, Point** and **Coordinate System**.

A new datum can be defined by a combination of existing datums and geometrical entities. For example, a new datum plane can be created as parallel (offset) to an existing plane or flat surface as demonstrated in Section 4.4.3. An axis can be a result of two intersecting planes, or it can be defined as through a point and perpendicular (normal) to a plane.

5.5.1 *Creating a datum plane through an axis*

The **Plane** ($\square$) command creates a new datum plane by selecting a combination of geometrical references that unambiguously defines a plane in the 3D space. The combination can be a single or multiple set of references from the following types:

- Parallel — Select a datum plane of a flat surface;
- Through — Select any of the following: (axis, edge and curve) or (point or vertex) or (plane) or (cylinder);
- Normal — Select any of the following: (axis) or (edge) or (curve) or (plane).

The next exercise will illustrate how to define a datum plane through an axis needed to create an angled hole in the BRACKET part.

(1) Open the BRACKET.PRT model.
(2) Move the cursor slowly above the right spindle hole to reveal the datum labels and click on axis **A_1** (Figure 5.12) to select.
(3) With the axis selected, click on **Plane** (⬜) icon in the mini menu or click on the **Plane** icon from the **Model** ribbon. A new datum plane, through the **A_1** axis, will be previewed and the **Datum Plane** dialogue window will open.
(4) Press (CTRL + Hold) and click on the vertical **FRONT** plane to add a reference. These two references fully define the new plane. Notice that the **OK** icon is now available (Figure 5.13).

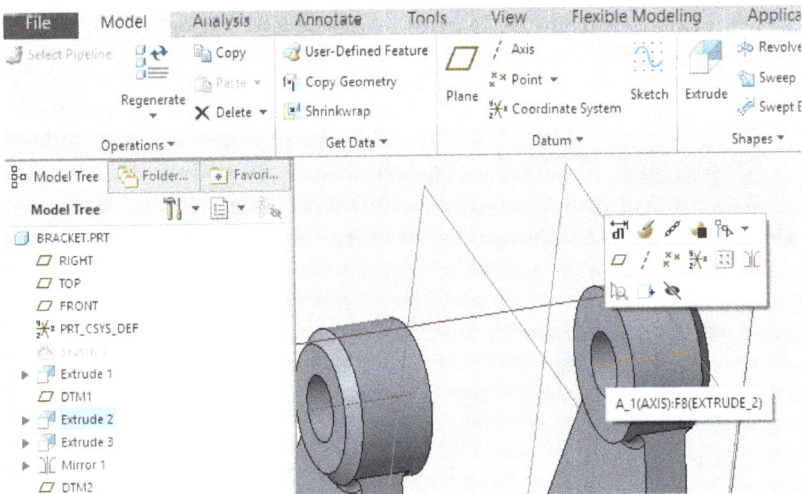

Fig. 5.12. Axis selection and the mini menu.

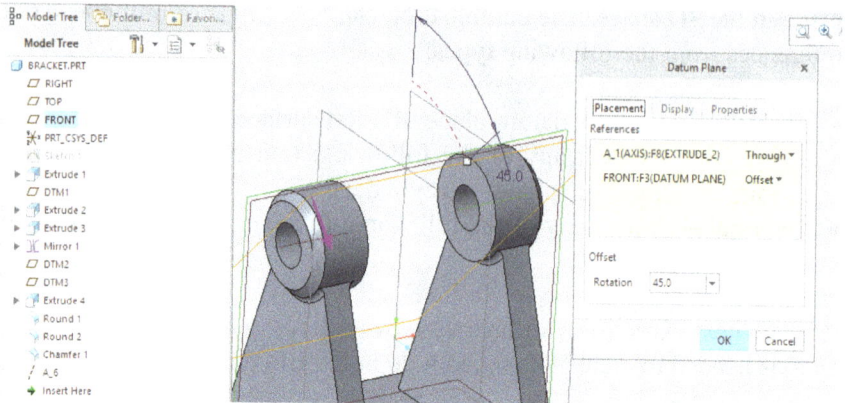

Fig. 5.13. Creating a datum plane through an axis.

(5) Drag the angle handle (small white square) to modify or type in **45** degrees angle in the **Rotation** dialogue box (Figure 5.13).
(6) Click on **OK** (**Datum Plane** dialogue window) to close the feature.

5.5.2 *Creating a datum axis as intersection of two datum planes*

(7) Select **DTM 1** and then click on the **Axis** (✓ Axis) icon from the mini menu or from the **Datum** group. A new datum axis, normal to **DTM 1**, will be previewed, as shown in Figure 5.14. Press (CTRL + Hold) and click on the last **DTM 4** datum plane to add another reference. The new axis is now shown at the intersection of the selected datum planes. Notice that the **OK** icon in the **Datum Axis** dialogue window will be available when sufficient references are selected.
(8) Click on the **OK** (**Datum Axis** dialogue window) to close the feature.

5.5.3 *Creating a datum axis through a cylindrical surface*

(9) Click on **Axis** (✓ Axis) icon in the mini menu or in the **Datum** group to start the command. Select the cylindrical surface of the corner round as shown in Figure 5.15. A new datum axis through the centre of the cylindrical surface will be previewed.
(10) Click on the **OK** icon (**Datum Axis** window) to close the feature.

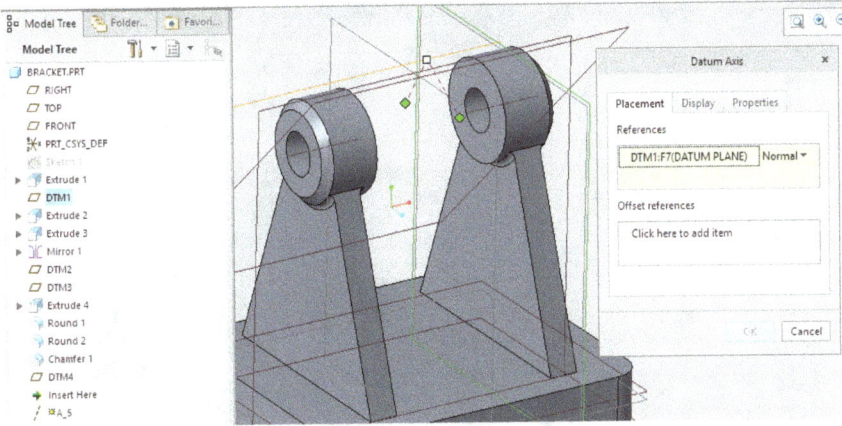

Fig. 5.14.   Creating a datum axis as an intersection of two datum planes.

Fig. 5.15.   Creating an axis through a cylindrical surface.

5.6 Creating Hole Features in the Bracket Part

(11) The next step is to create <u>an angled hole</u> in the spindle support. Select the axis **A_5** (the axis of the round cylindrical surface) and then click on **Hole** (Hole) icon to start. Press (CTRL + Hold) and click on the cylindrical (spindle support) surface (as Placement references). A preview of the hole appears, as shown in Figure 5.16.

(12) Move the cursor to the **Hole** dashboard, click on the arrow to open the pull-down options and select **Drill to selected**… as hole depth.

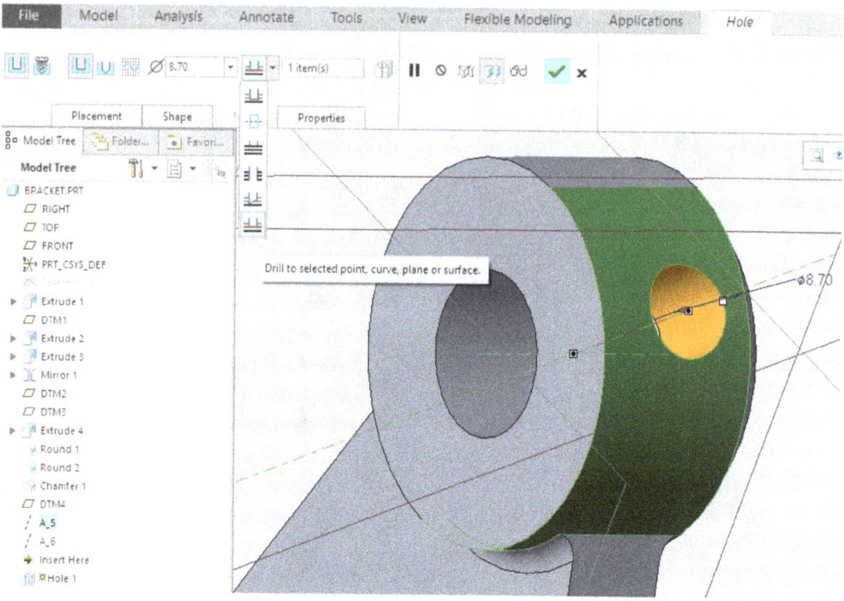

Fig. 5.16. Creating an angled hole through an axis.

Move the cursor and select the central axis as the hole depth reference (Figure 5.16).

(13) Enter **3** mm in the diameter size dialogue box and then click on the green tick (✓) icon to close the **Hole** dashboard.

(14) With the hole feature still selected (select if necessary), click on the **Mirror** icon and select the middle datum plane (**RIGHT**) to mirror the hole (in the symmetrical spindle support).

(15) Click **Save** (Save) to save the model.

(16) Continue with the BRACKET modelling by creating <u>counter bore holes</u> in the part base as follows: Select the axis through the cylindrical surface, then click on the **Hole** (Hole) icon to start. The **Hole** dashboard opens. Press (CTRL + Hold) and click (LMB) on the top flat surface to add another reference for hole placement. A preview of the hole will appear in the Graphics area (Figure 5.17).

(17) In the **Hole** dashboard, click on the **Use standard hole profile** ( ) icon (Figure 5.17). Next, click on the counter bore icon ( ) to specify the hole type.

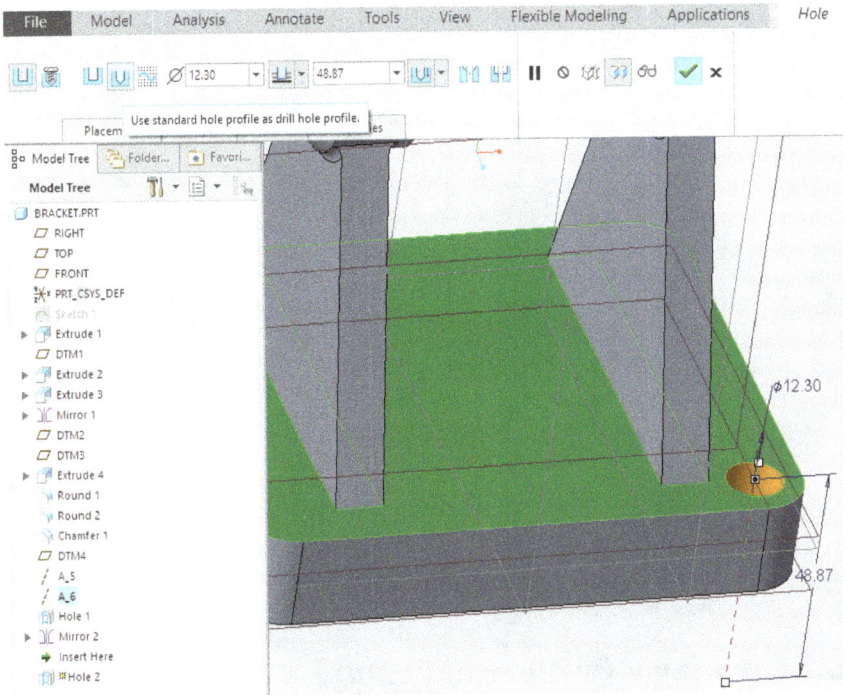

Fig. 5.17.   Creating a counter bore hole through an axis.

(18) Click on the **Shape** tab to reveal the counter bore dialogue box
(Figure 5.18). Type in **12** and **6** mm for the counter bore diameter
and depth, **8** mm for the hole diameter and select **Through All**
option from the hole depth drop-down menu (click on the black
arrow).

(19) Click on the green tick (✔) icon to close the **Hole** dashboard.

(20) With the last feature still selected, click on the **Mirror** tool, and click
on the middle datum plane (**RIGHT**) to mirror the hole. If the result
is as expected, then close the **Mirror** dashboard.

(21) Press (CTRL + Hold) and click (LMB) on the previous hole and the
mirrored feature. Click on the **Mirror** icon again and select the
FRONT plane as mirror datum plane. Click on the green tick to
close the feature. Four symmetrical holes will be created.
Note: All mirrored holes are depended (<u>Children</u>) on the very first
hole, which is their <u>Parent</u>. Therefore, to modify all holes, the user

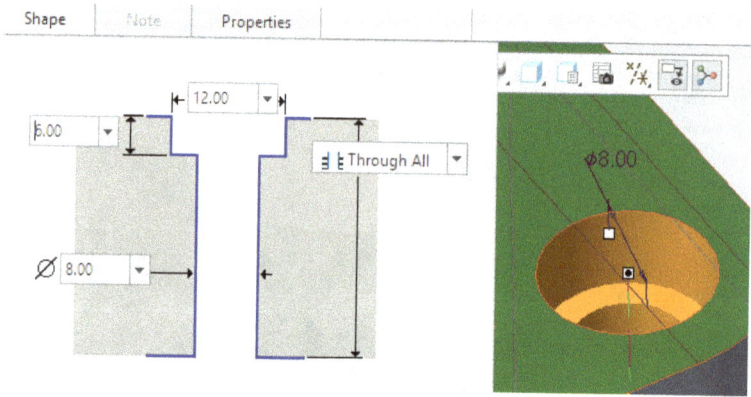

Fig. 5.18.   Shape tab (**Hole** dashboard).

needs to modify only the <u>Parent</u> hole, and the rest will assume the changes.

(22) Click **Save** (Save) from the toolbar to save the model.

5.7 Creating a Profile Rib Feature

There are several ribs in the Bracket design as shown in Figure 4.1. They could all be created with the **Extrude** command. However, a more elegant way to create ribs with complex geometry is by using a special command called **Rib** from the **Engineering** group. The command has two sub-commands — **Trajectory Rib** and **Profile Rib** — that can be activated from the **Rib** pull-down menu. (Click on the black arrow next to the **Rib**.)

The **Profile Rib** command is similar to the **Extrude** command and requires a sketching plane, a 2D sketch of the profile and rib thickness.

(23) Select the **FRONT** datum plane and then click on the **Profile Rib** (Trajectory Rib) icon to activate the command, i.e. **Model > Engineering > Profile Rib**.

(24) The Sketch will open for the user to draw the rib profile. Click on the **Sketch View** ( ) to align to the screen.

(25) Click on **References** and select all adjacent surfaces that will connect to the rib, i.e. top of the base and the two inner support surfaces, as shown in Figure 5.19. The references will appear as blue dashed lines. Click on **Close** in the **References** window.

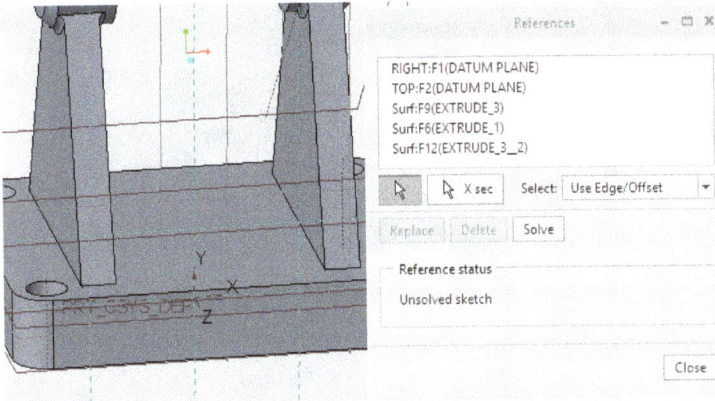

Fig. 5.19.   Profile Rib command — references selection.

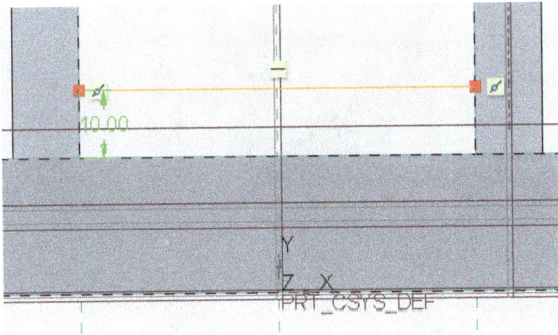

Fig. 5.20.   Profile Rib sketch.

(26) Pick the **Line Chain** tool and sketch a single horizontal line snapping the ends to the two vertical references. Click MMB to stop. Note that the line will create an open contour which is acceptable for the **Rib** command (Figure 5.20). Type **10** mm as distance dimension.

(27) Click on the **OK** icon (✔) to save, exit the **Sketch** and return to the **Profile Rib** dashboard.

(28) The magenta arrow indicates the direction of the rib. Click on the arrow to switch the direction towards the base. The preview will show the rib connected to the supports and the base (Figure 5.21). Type **8** mm as rib thickness.

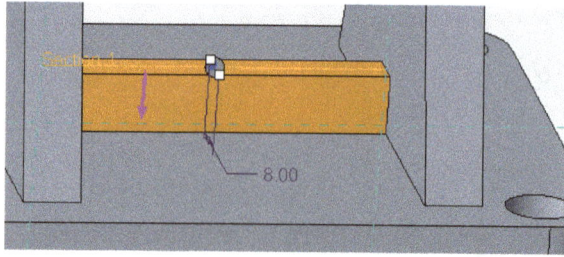

Fig. 5.21.   Profile Rib feature — thickness direction.

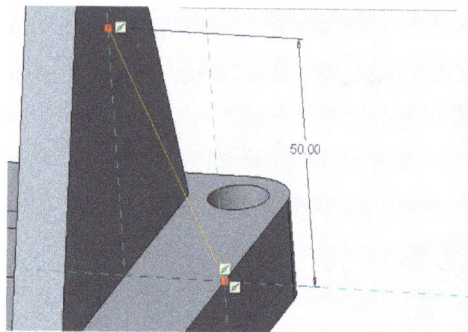

Fig. 5.22.   The triangular right side rib.

Notice that the rib thickness is symmetrical against the sketching plane. To toggle the thickness direction to one side or to the other, click on the arrow icon ( ) in the **Rib** dashboard.

(29) Click on the green tick ( ) to accept and close the **Profile Rib** feature. If a fundamental error has been made, then click on the Cancel icon ( ) to abort and start again.

(30) Repeat the **Profile Rib** command to create the triangular rib on the right side support (see Figure 5.22). Select the following three surfaces as references: the top and the side vertical surfaces of the base as well as the support vertical surface. Sketch a single line connecting to the right edge and to the vertical side surface. Use **50** mm dimension as height (Figure 5.22) and **8** mm rib thickness.

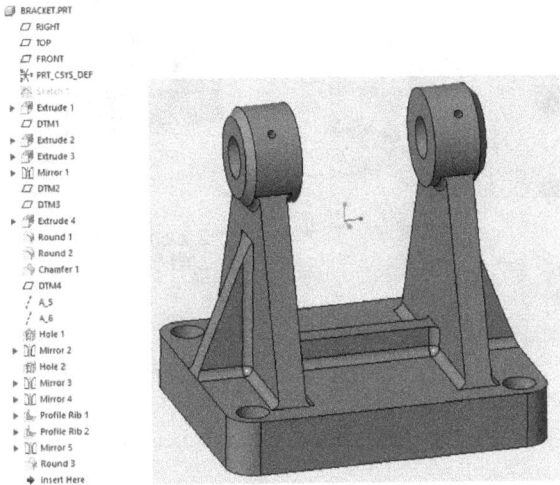

Fig. 5.23. The BRACKET part.

(31) Use the **Mirror** command to mirror the previous rib to the left.
(32) Use the **Round** command to add rounds with **3** mm radius between the ribs, supports and the base as shown in Figure 5.23.
(33) Save the model, **File > Save** (Save).

The final BRACKET part should look as shown in Figure 5.23.

5.8 Exercises

Exercise 1
Create a solid part with uniform wall thickness of **10** mm as shown in Figure 5.24. Draw the plan view as initial sketch. Apply construction centrelines and symmetric constraints to promote part symmetry. Use **Extrude** commands to add or remove material. When sketching, include references from the previous features and use the **Project** tool (Sketch) to copy existing edges.

Exercise 2
Create a solid part as shown in Figure 5.25. Use **Revolve, Hole** and **Pattern** features.

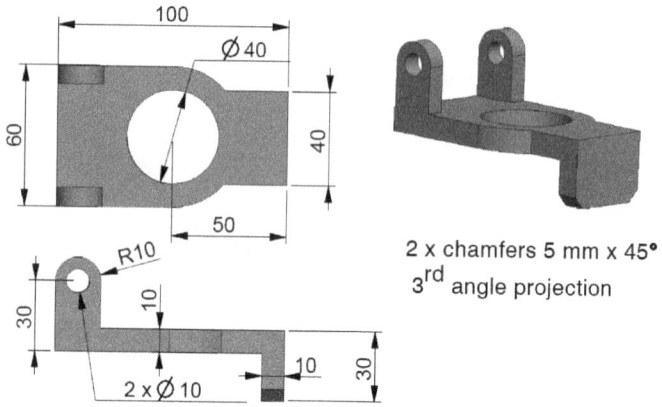

100

Ø 40

60

40

50

R10

10

30

10

30

2 x Ø 10

2 x chamfers 5 mm x 45°
3rd angle projection

Fig. 5.24.   Solid part — Exercise 1.

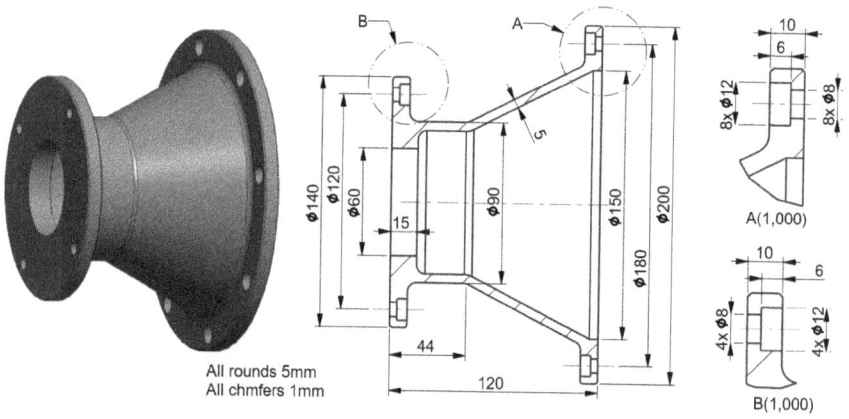

B

A

10

6

8x Ø12

8x Ø8

Ø140

Ø120

Ø60

Ø90

Ø150

Ø200

Ø180

5

15

44

120

A(1,000)

10

6

4x Ø8

4x Ø12

B(1,000)

All rounds 5mm
All chmfers 1mm

Fig. 5.25.   Solid part — Exercise 2.

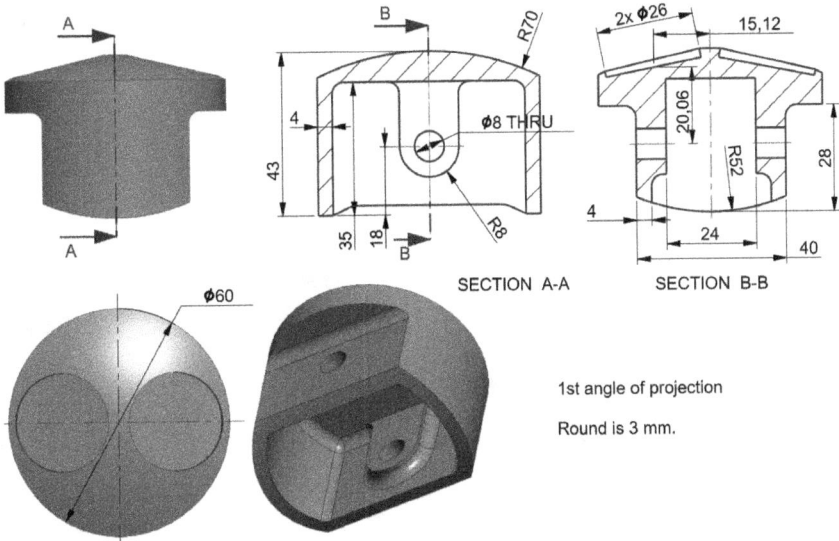

Fig. 5.26.   Solid part — Exercise 3.

Exercise 3

Create a solid part as shown in Figure 5.26. Start with a **Revolve** feature to create the main shape, then use **Extrude** to add or remove material. Create an axis through the point with coordinates (**20.06** and **15.12** mm) and also normal to the plane as shown section in B-B. Create an angled datum plane through that axis in order to sketch the two **26** mm diameter cut-outs (piston top). Apply **Mirror** command for symmetrical details.

Chapter 6

Creating a Part Suitable for Plastic Injection Moulding

6.1 Introduction

Injection Moulding (IM) is widely used manufacturing technology for producing plastic parts in high volume at low cost. It is essential to know how the IM technique works in order to understand the design of parts suitable for this type of manufacture. The IM process involves heating and melting a thermoplastic material that is injected into a metal mould tool under high pressure. The plastic solidifies inside the mould, assuming the mould shape, and then the mould tool opens allowing the part to be extracted (ejected).

Some of the rules that a CAD designer has to follow when modelling parts for this technology are the following: design the part with a shape that allows for easy extraction from the mould tool, maintain uniform wall thickness, round the corners to facilitate the low of molten plastic, avoid concentration of thick areas that might develop sink marks on the part visible side and create drafts that will help with the part extraction from the mould. Creo has a number of features and commands that facilitate the part and mould design.

In this lesson, the reader will apply the knowledge and skills gained from the previous chapters and learn new commands in order to create a 3D model suitable for plastic IM, adopting the best industrial practices.

Aim:
To develop modelling skills required to create parts with correct attributes for efficient IM manufacture.

Outcomes:
At the end of this lesson, you should be able to:

- Create a part model with uniform wall thickness using the **Shell** feature;
- Design webs using the **Extrude** feature with **Thicken Sketch** option;
- Create a base sketch for the **Profile Rib** feature;
- Create a **Group** of features;
- Dynamically modify model dimensions;
- Learn how to apply the **Sweep** feature;
- Use the **Draft** feature and apply drafts correctly in order to make the part suitable for IM;
- Consider and use preferred sizes in the choice of dimensions;
- Create a section view and analyse the model geometry;
- Consider the application of additional design improvements and factors that influence the degree of draft angle;
- Create model analysis.

6.2 Creating the Main Shape of a Plastic Cover

The tasks described in the following sub sections will teach the reader how to model a part (Plastic cover) suitable for IM.

The modelling process will incorporate solid extrude and shell features, bosses, suitable ribs to the underside, draft feature and rounds to blend the corners and smooth the wall thickness variation.

(1) Start Creo (unless it is already running) and set Working Directory: **File > Select Working Directory** > C:\USER\CREO_PRACTICE.

(2) Start a new part: Click on the **New** (⬜) icon from the Quick Access Toolbar or click on **File > New**.

(3) Type the name PLASTIC_COVER in the pop-up **New** window (Figure 6.1). Keep all options default (Part mode), untick the **Use default template** slot and click on the **OK** (OK) icon to accept.

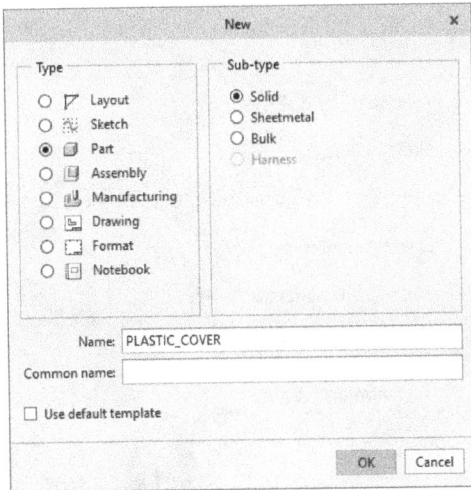

Fig. 6.1. The New part window.

(4) Select an appropriate System of Units in the **New File Options** window, i.e. select **mmns_part_solid** for Metric units system (metre, Newton, second) and click on the **OK** (OK).

🖉 The Part mode window will open with the standard model ribbon interface. The Model Tree contains the three main datum planes and the absolute coordinate system.

🖉 Please notice the message area across the bottom of the main window. Various messages and prompts are posted here to help the user provide correct inputs. The window can be scrolled to view a complete history of the current session. KEEP WATCHING THE MESSAGE AREA.

The main part shape will be created as an **Extrude** feature with dimensions **150 × 60 × 35** mm, as follows:

(5) Click on the **Extrude** icon (▱) from the ribbon to start.
(6) Click on the **Placement** tab (**Extrude** dashboard) to open the panel (Sketch), and then click on the **Define** icon. The **Sketch** dialogue window opens (Figure 6.2). In this window, set up the corresponding references for the **Sketch Plane** and **Sketch Orientation**. Select the **TOP** datum plane, from the Model Tree or Graphics area, as the

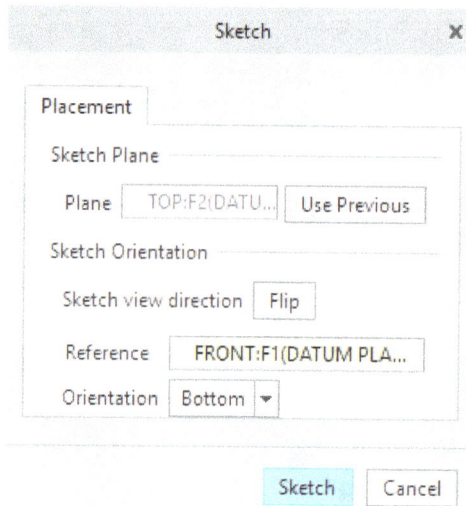

Fig. 6.2.   The Sketch window.

sketching plane. **FRONT** datum plane is selected automatically as default **Reference** plane. If the selection is acceptable, click on **Sketch** to proceed to the Sketcher.

The selected planes are highlighted in green and blue. The **Sketch view direction** arrow (in magenta) can be flipped. Alternatively, select another **Reference** plane for a different orientation. Note that the corresponding datum names and parameters are set in the window slots (Figure 6.2).

(7) Select **Sketch View** ( ) from the Graphics toolbar to align the sketching plane to the screen.

The user can set up the sketching plane to align to the screen auto-matically by modifying the value of the parameter **sketcher_starts_in_2d** to **yes**. To do this, select **File > Options > Configuration Editor.** Click on the **Find** icon (at the bottom of the window) and type *sketcher_starts* in the **Type keyword** slot. Click on **Find now**, select **sketcher_starts_in_2d** and in the **Set value** slot change the value from **no** to **yes**. Click on **Add/ Change**, then **Close**, and **OK.** Select **Yes** and **OK**, to save to a new configuration file. Next time when you start the Sketcher, you do not need to do the previous step.

(8) Start drawing the section. Select **Centreline** (┊) tool from the **Centreline** types drop-down menu (**Sketching** group) or right-click and select **Centreline** (┊) from the **Sketch tools** mini menu. Position the cursor to snap at the horizontal reference line (dashed line) and click to place the first point. Move the cursor away, snap it again at the same reference and click to place the second point.

(9) Continue drawing a second centreline. Position the cursor on the vertical reference line and click for the first point, then move the cursor and click on a second point on the vertical reference. (Click on **Centreline** (┊) tool again if the command has been interrupted).

(10) Click on the **Corner Rectangle** (▢) icon from the **Rectangle** types drop-down menu, the **Sketching** group, or (RMB + Hold) start **Sketch Tools** and select **Corner Rectangle** (▢). Position the mouse cursor at the upper left quarter and click on a point to start the top left rectangle corner. Move the cursor down to the lower right rectangle corner and click on a point to finish. Click the MMB to stop the **Corner Rectangle** command (Figure 6.3, left). When moving the cursor, try to find a symmetric position for the second point. However, if the rectangle vertices are not symmetric, then apply symmetric constraints against the vertical and horizontal centrelines. To do so, click on the **Symmetric** (⊣⊢ Symmetric) icon and, following the message line prompt, first select the centreline and then the two vertices that should become symmetric. Click the MMB to stop. The sketch should look similar to Figure 6.3, right.

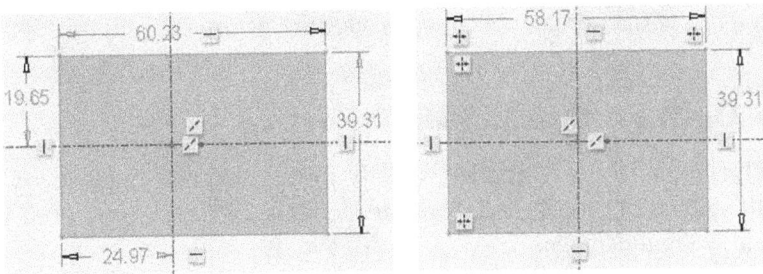

Fig. 6.3. Rectangle — without symmetrical constraints (left); with symmetrical constraints (right).

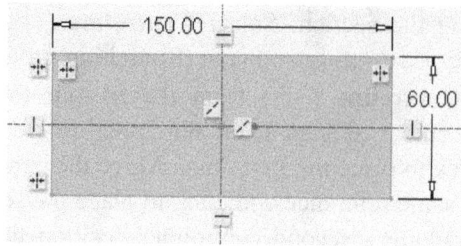

Fig. 6.4.   Rectangle — final sketch with dimensions.

Notice the **Vertical, Horizontal** and **Symmetric** constraint icons.

(11) The rectangle size is probably incorrect. To edit, click on the **Select** (↖) and double-click on each blue dimension until it highlights in the slot. Type **150** mm for the horizontal and **60** mm for the vertical dimensions and press ENTER. The result is shown in Figure 6.4.

(12) Click on the **OK** icon (✓) to save, exit the Sketcher, and return to the **Extrude** dashboard.

(13) In the **Extrude** dashboard apply **30** mm depth in one direction as shown in Figure 6.5, left.

(14) If the result is acceptable, click on the green tick (✓) to save and complete the feature. The view returns to the main Creo desktop view showing the extruded part (Figure 6.5, right).

Remember to provide all necessary inputs in the feature dashboard. The indication for it is when the green tick (✓) icon is available. If not, then try to correct or use cancel (✗) and start again.

6.3 Draft Feature and Draft Surface

Drafts are features applied directly to the existing 3D geometry to create angled or sloped surfaces that facilitate part extraction from the mould without damage.

The command that creates drafts on a single or several surfaces is called **Draft**. It can be applied only on extruded surfaces to create an angular deviation from the normal to the parting plane direction. A draft

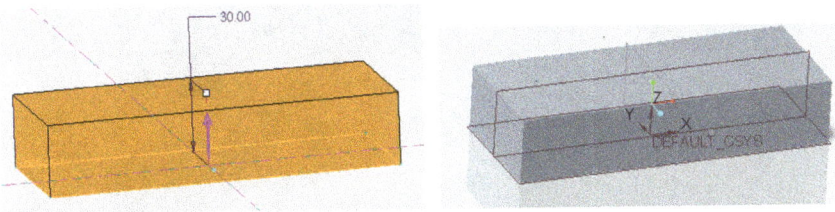

Fig. 6.5.   Extrude feature — preview (left) and after closing the feature (right).

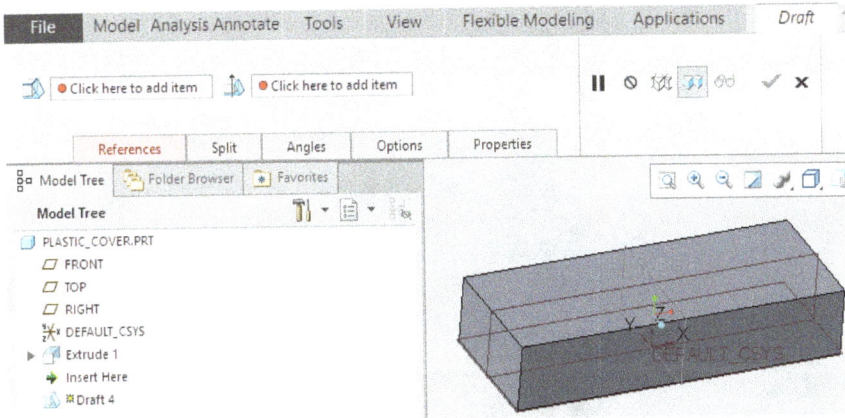

Fig. 6.6.   Draft feature dashboard.

can be thought of as being a 'door'. The draft surface is the door face, and the draft hinge is the door hinge. The <u>pull direction</u> is the normal (vector) to the <u>draft hinge</u>. Draft command can apply angles from **0** to **90** degrees. However, in plastic IM, the draft angles are limited to small values from **0.5** to **5** degrees. Drafts of **2** degrees will be applied to the PLASTIC_ COVER part as follows:

(15) Click on the **Draft** ( ) icon to open the draft dashboard (Figure 6.6). Click on the **References** tab to expand (Figure 6.7).

(16) Click on the **Draft Surfaces** slot to activate. Press (CTRL+ Hold) and then click on all four peripheral surfaces to select them. They should be highlighted in green colour, as shown in Figure 6.9.

To check the selected surfaces, click on (Details...) icon and open **Surface Sets** window (Figure 6.8). Notice that the list has four surfaces.

| References | Split | Angles |

Draft surfaces
● Select items   Details...

Draft hinges
● Click here to add item   Details...

Pull direction
● Click here to add item   Flip

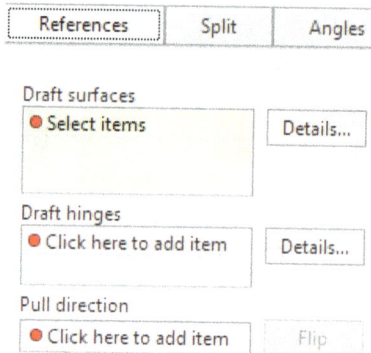

Fig. 6.7.   Draft — References tab.

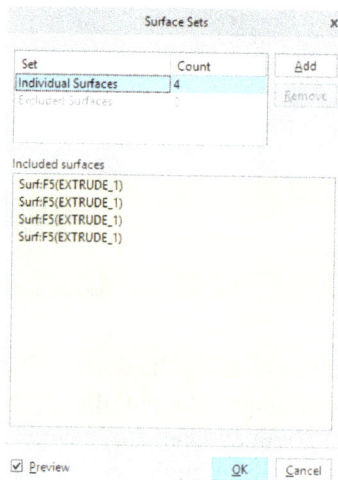

Surface Sets   ✕

Set	Count	Add
Individual Surfaces	4	
Excluded Surfaces		Remove

Included surfaces
Surf:F5(EXTRUDE_1)
Surf:F5(EXTRUDE_1)
Surf:F5(EXTRUDE_1)
Surf:F5(EXTRUDE_1)

☑ Preview   OK   Cancel

Fig. 6.8.   Surface Sets window.

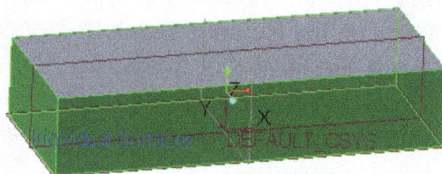

Fig. 6.9.   Draft surfaces.

If a mistake has been made, press (RMB + Hold) to open a menu and then LMB click on **Remove** or **Remove All** to unselect. Click on **OK** to close.

6.3.1 *Draft hinges and pull direction*

The <u>Draft hinge</u> and <u>Pull direction</u> are datum planes or flat surfaces selected as draft references. Also, the <u>Draft hinge</u> can be a curve chain. The draft surfaces pivot against their intersection with the draft hinge.

The <u>Draft hinge</u> does not change its geometry when a draft is applied. For example, the bottom surface in Figure 6.10 will remain the same, but the top surface will be smaller (or larger) due to the draft angle.

The <u>Pull direction</u> is the normal (vector) to the selected surface (datum plane) that the part would follow when 'pulled' out of the mould. The draft angle is measured between the surfaces to draft and the pull direction surface. By default, the <u>Pull direction</u> (References tab) is the same as the <u>draft hinges</u>.

(17) Click inside **Draft Hinges** slot (**References** tab) to activate it for selection, then select the draft hinge on the model. Click on the bottom surface of the part to select. The **Pull Direction** slot will assume the same surface name. Note the pink arrow indicating the <u>pull direction</u>. You can practice by making various selections for draft surfaces, draft hinges, and pull directions. If the selection set is correct, a preview of the draft will be shown (Figure 6.10). Notice the names of selected surfaces in the **References** slots.

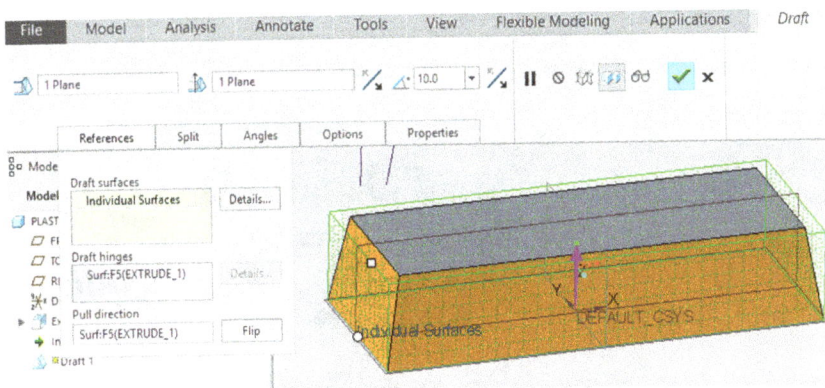

Fig. 6.10.   Draft feature — preview.

(18) By default, the draft angle is **1** degree, and often it is difficult to judge whether the surfaces are in the correct (positive draft) or incorrect direction. Type a larger value, for example **10** degrees, in the dashboard angle slot ($\angle$ 10.0 ▼) or in the Graphics area, in order to exaggerate the view. If the direction is wrong, then type **-10 (minus 10)**. Alternatively, click and drag (LMB + Hold) the draft handle (white rectangle) to apply an angle. When you are satisfied with the direction, type the correct value, i.e. **2** or **3** degrees.

Notice that the bottom surface (draft hinge) is not affected by the draft. However, the top surface size is smaller what it was originally before the draft. If you flip the Pull direction arrow, then the top surface will be larger.

(19) If the result is satisfactory, then click on the green tick (✓) to complete the draft feature. (Figure 6.11).

Look at the Model Tree pane (left on screen). As each feature is completed, it is appended to the Model Tree. The next feature will be created after the last feature listed on the Model Tree.

Always apply **Round** or **Chamfer** features after **Draft** feature. Drafts will not work on surfaces with adjacent rounds or chamfers.

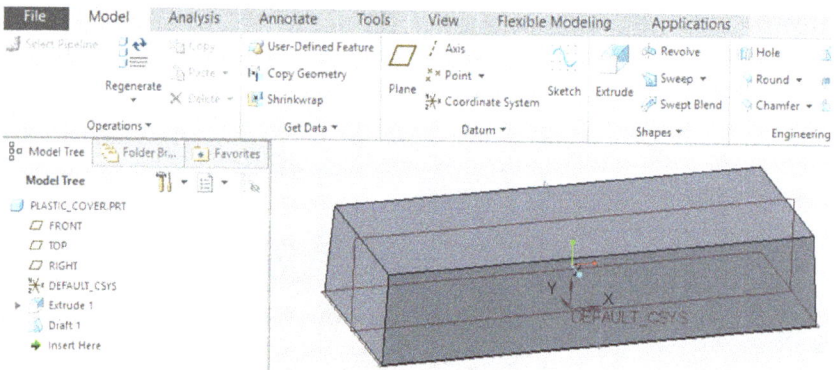

Fig. 6.11.   Draft feature completed.

6.4 Creating Rounds

The next step is to apply rounds to all edges of the PLASTIC_COVER, excluding the four edges of the open face of the shell.

(20) Click on the **Round** (⏣) icon to open the command dashboard.
(21) Press (CTRL + Hold) key and click on the four vertical and four horizontal edges. A preview of the rounded edges is shown in Figure 6.12. The selected horizontal edges should belong to the surface that has reduced size due to the draft.
(22) Set the rounds radius to **4** mm.
(23) Click on the **Sets** tab to reveal the drop-down panel (Figure 6.13).

ⓘ **Round** is a powerful command that creates rounds with constant or variable radius. The round geometry can be either a circle or conic. The default settings shown in Figure 6.13 are a circular round with constant radius. To create a round with variable radius, move the cursor over the **Radius** slot (**Sets** tab) and click (RMB + Hold), then from the menu that will appear, select **Add radius**. A new radius entry will be added to the table where you can type a new value. Every radius is associated with a **Location** (a vertex or a point on the selected edge). Explore the options for more complex rounds available in the **Sets** tab (Figure 6.13).

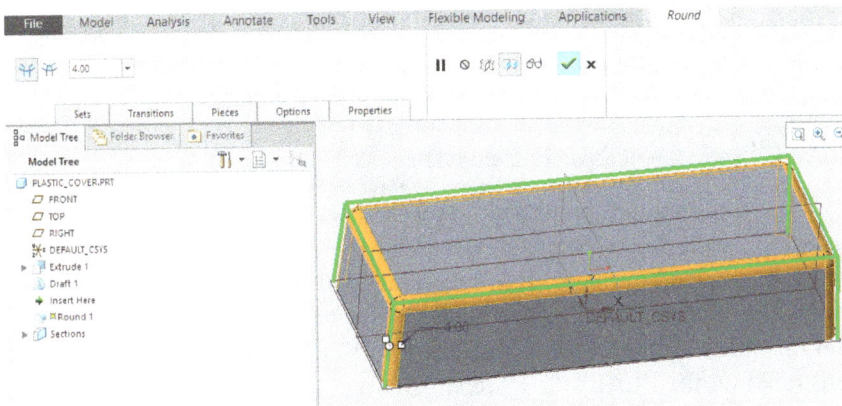

Fig. 6.12.   Round dashboard — rounds preview (8 edges selected).

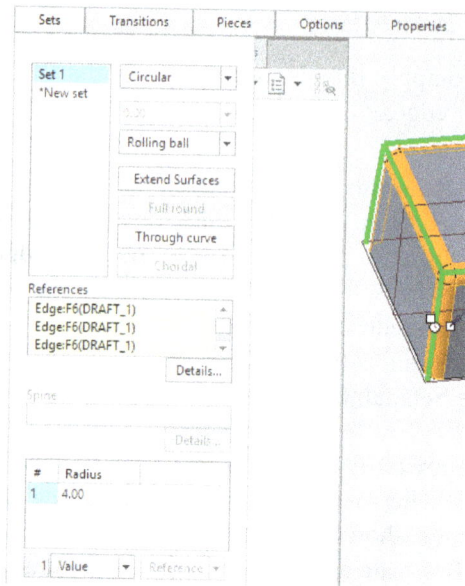

Fig. 6.13.   Sets drop-down panel (Round dashboard).

(24) If the preview is as expected, click on the accept icon (✔) to close the **Round** command.

(25) The view returns to the main **Model** ribbon interface showing the part with rounds. Note the new round feature in the Model Tree.

(26) Click on **File** > **Save**. Save less frequently if you feel more confident.

6.5 Creating a Shell Feature

ⓘ The **Shell** feature (command) creates a shell with uniform wall thickness following the external part geometry by removing material from one or more selected surfaces. It is recommended to use the shell command after all rounds and drafts have been applied to the model. Shell feature is very useful for creating shell models for IM parts, and this will be demonstrated on the PLASTIC_PART to create a shell of **3** mm thickness.

(27) Click on the **Shell** icon (▣) and open the dashboard (Figure 6.14).

(28) Select the bottom surface (the surface without rounds). A preview of the shell is shown in the Graphics area (Figure 6.14).

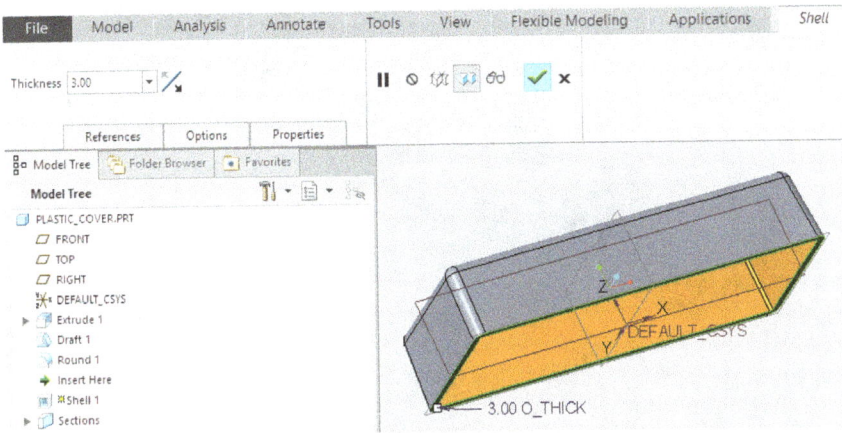

Fig. 6.14. Shell dashboard — preview after selection of the surface to be removed.

(29) Click on the **Thickness** icon (Thickness 3.00) and enter **3** mm.

(30) If the result is ok, then click on the green tick (✓) to accept and close the dashboard.

The **Shell** command generates a set of offset surfaces from all external geometries to create the internal surfaces. Small rounds can be problematic if the shell thickness value is equal to or larger than the smallest round radius. In such cases, the **Shell** will fail because of overlapped geometry.

6.6 Creating Webs — Extrude with Thicken Sketch Option

After shelling the part, the next steps are to make the cross webs inside the plastic cover. The webs will be created as thin extrusion features with a thickness of **1.5** mm. The section (sketch) will be drawn on a plane that is offset from the part's top face. The webs must fully merge with all internal surfaces without any voids.

(31) First, an offset datum plane must be created. Click on the **Plane** icon ( ) from the **Model** ribbon, **Datum** group.

(32) Select the bottom surface (**TOP** datum plane), drag the offset handle (small white square) upwards to **15** mm (or type the value in

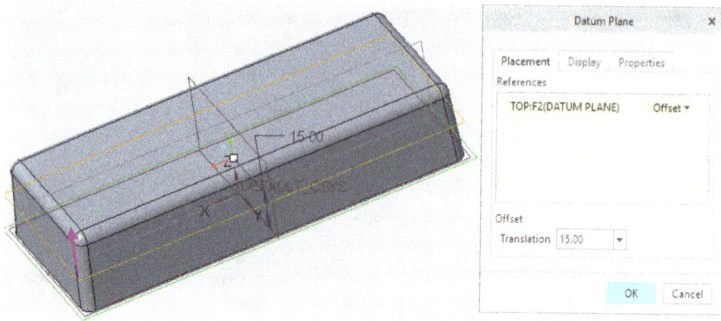

Fig. 6.15.   Offset datum plane.

Translation slot) from the **TOP** (Figure 6.15) and click on **OK**. This offset datum plane will be used to sketch the web sections.

(33) Click on the last datum plane (from the Model Tree or Graphics area), unless it is already selected, and then click on the **Extrude** (⌐) icon. (The **Extrude** can be invoked from the mini tool bar that will appear above the cursor when you click on a feature.)

(34) The **Sketch** ribbon will open. (Click on the **Sketch View** (🖼) icon to orient the sketching plane parallel to the screen if needed.)

(35) Click on the **Sketch Setup** (✏ Sketch Setup) icon (top left in the ribbon), then click on **Flip** from the **Sketch** dialogue window to change the sketching orientation without leaving the Sketcher environment. The screen should look as in Figure 6.16. Click on **Sketch** to continue.

(36) Click on the **Line Chain** tool and start drawing a horizontal line. Move the cursor to the left until it snaps to the part inner edge, then move the cursor down until it also snaps to the horizontal reference line and click the LMB to create the line first point. Move the cursor horizontally until it snaps to the opposite inner vertical edge (Figure 6.17) and click to create the line's second point. Click MMB to finish the sketch. Note that the sketched line does not have a dimension because it is attached to the existing geometry (references).

(37) Click on the **OK** (✔) icon to close the **Sketch**. Generally, the **Extrude** command cannot create a solid feature from an open section, a line in this case. However, the **Extrude** has an option — **Thicken Sketch** — that can. Click on the **Thicken Sketch** icon (⌐) and preview the result, as shown in Figure 6.18. The thickness

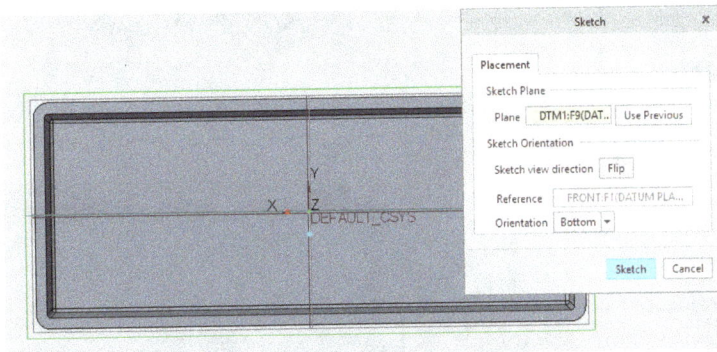

Fig. 6.16.   Changing of the sketch orientation.

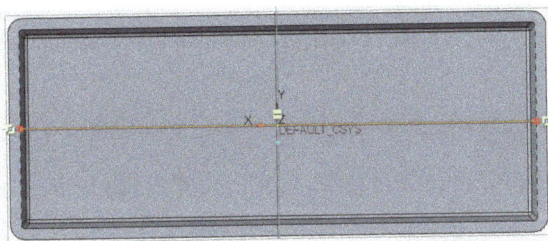

Fig. 6.17.   Selecting the points of the line to snap to the vertical inner edges and the horizontal reference (reference datum plane).

of **1.5** mm can be entered directly in the <u>thickness value</u> slot. The change direction icon (⟋ₓ) can toggle the thickness to **one side**, the **other side** or **both** sides of the sketched line. Click on this icon several times until the thickness appears in both directions.

The rib thickness of **1.5** mm has been chosen as 50% of the main wall thickness (**3** mm) to prevent sink marks. The general rule for a rib thickness is that it should be 40%–60% of the main wall thickness. Note that if a draft is applied at the rib at both sides, the rib thickness at the base might increase.

(38) Switch **Extrude** direction towards the part inside and select **Extrude**

up to next … icon (⊟) as the extrude depth option (Figure 6.19).
(39) Click on the green tick (✓) to close the **Extrude**.

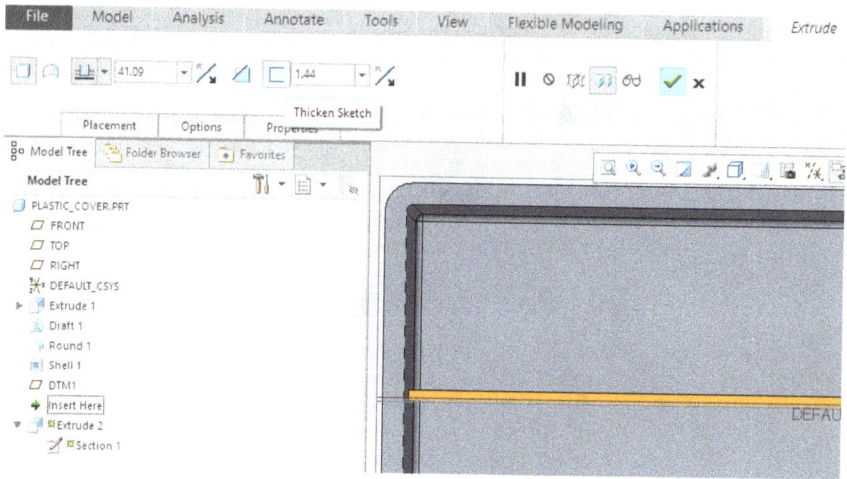

Fig. 6.18.   Thicken Sketch option within Extrude command.

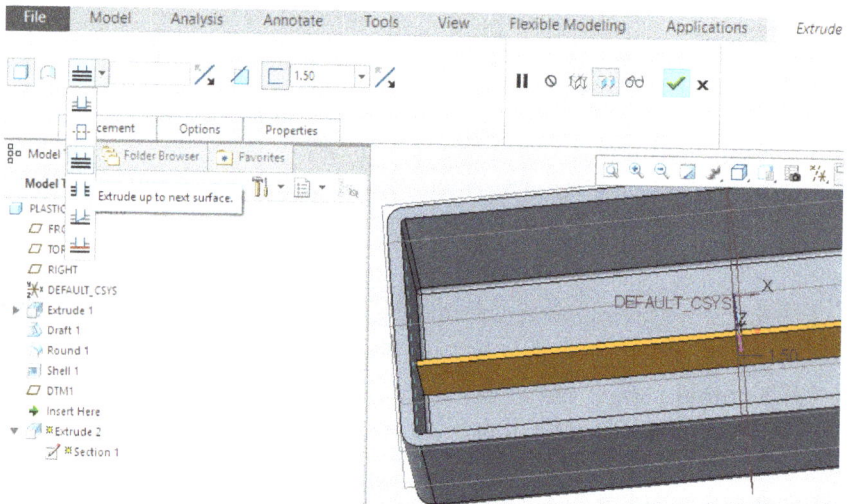

Fig. 6.19.   Extrude up to next surface as depth option (Extrude).

6.6.1 *Creating a datum plane for extrude* (*with thicken sketch option*)

Before continuing with the ribs, create another datum plane that will be used as a reference for sketching the line for another rib.

(40) Click on the **RIGHT** datum plane and start the **Plane** ($\square$) command. Drag the offset handle upwards to **30** mm (**Translation**) from **RIGHT** and click on **OK**. A new **DTM 2** will be created.

All Creo commands work either by starting the command first and then selecting the references, or selecting the reference first and then activating the command itself, as demonstrated in the above step.

(41) DTM 2 will be the additional reference for the second rib. Select **DTM 1** as the sketching plane. Then start another **Extrude** feature (with **Thicken Sketch** option), activate **References** ($\square$) from the **Sketch** ribbon and select **DTM 2** as a reference for the sketch. Draw the vertical rib as in the previous steps, snapping to the top and bottom part inner edges and also to the vertical (**DTM 2**) reference. Use **Extrude up to next surface** ($\underline{\underline{\underline{\quad}}}$) as extrude depth and **1.5** mm thickness (Figure 6.20). Close the feature if the result is ok.

(42) Repeat the **Extrude (Thicken Sketch)** option, creating more ribs.

Do not sketch crossed lines or branched contours with this approach. The **Extrude** cannot resolve such geometry.

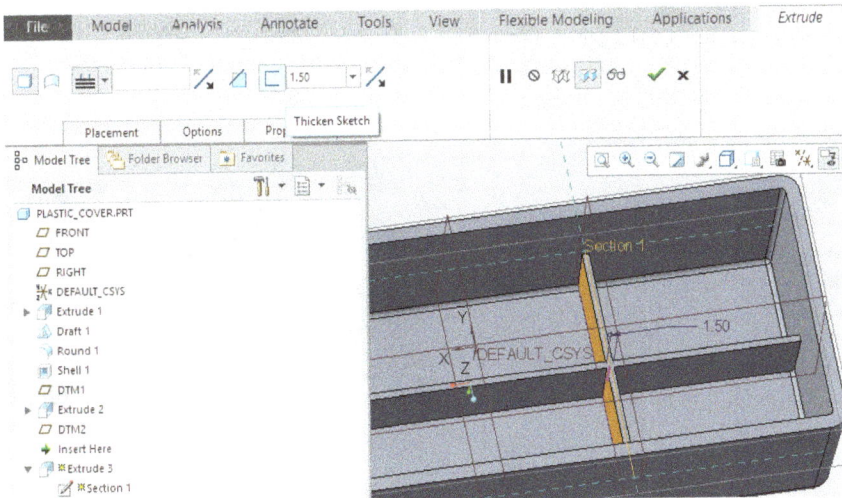

Fig. 6.20.   The vertical rib as Extrude (Thicken Sketch) feature.

6.7 Creating a Boss as Extrude Feature

(43) Select the **TOP** datum plane and click on the **Extrude** icon to start. Click on **References** (⬚), the **Sketch** ribbon, and pick **DTM 2** as a reference. Sketch a **Circle**, snapping the centre to the horizontal and vertical references. Enter **8** mm as circle diameter. Use **Extrude up to next surface** as depth (Figure 6.21) and close the feature.

6.8 Creating a Profile Rib Feature

The **Profile Rib** feature (Chapter 5, Section 5.7) will be used to create triangular ribs to support the boss.

(44) Select the **TOP** datum plane and then click on the **Profile Rib** (⬚ Profile Rib) icon, **Engineering** group, to start the feature. The **Sketch** will open to allow sketching of the rib profile.

(45) Select **References** (⬚), and click on all surfaces that connect to the rib. In this case, select the boss silhouette (cylindrical surface) and the top of the adjacent horizontal rib (Figure 6.22). The references are highlighted in blue dashed lines and appear in the **References** window. Click on **Close and continue**.

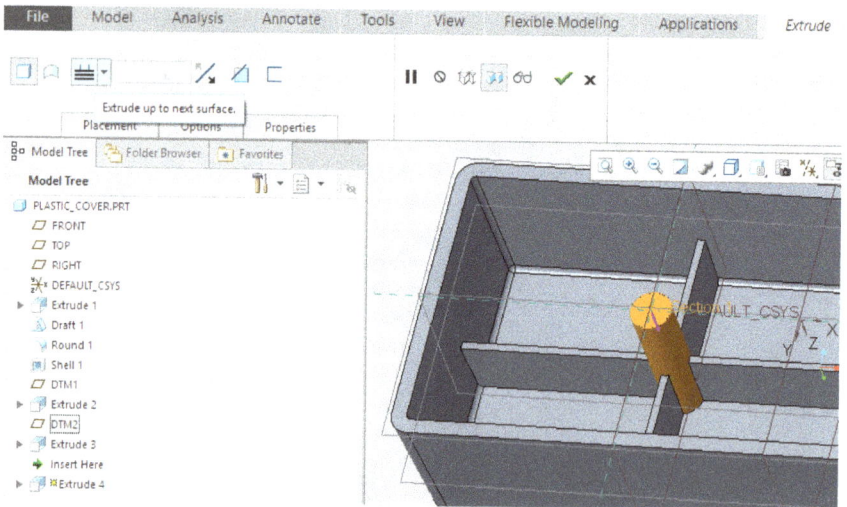

Fig. 6.21.   A boss created as extrude feature.

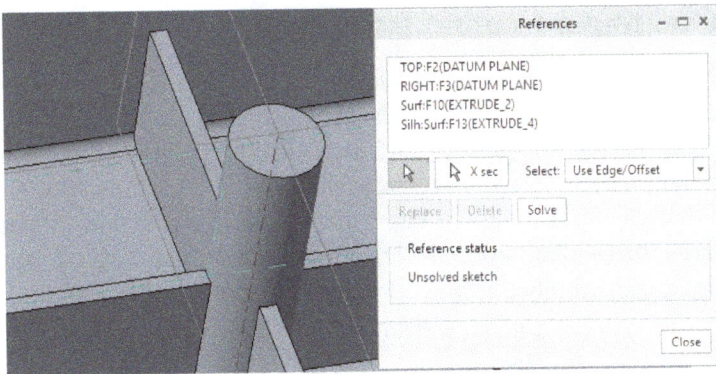

Fig. 6.22. Profile Rib — references selection.

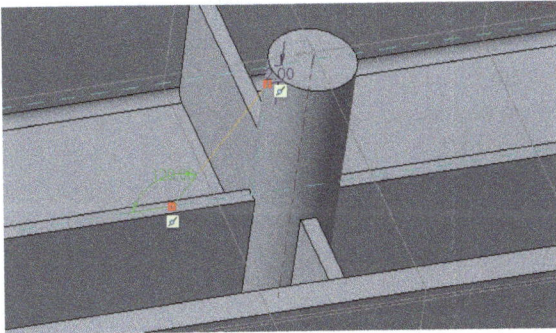

Fig. 6.23. Sketching Profile Rib.

(46) Click on the **Line Chain** tool and sketch a single line snapping the line points to the above references (Figure 6.23). Click MMB to stop. Enter **2** mm (distance from the boss top) and **120** degrees as dimensions. Alternatively, select a different dimensioning scheme.

(47) Click on the **OK** (✔) icon to close the **Sketch** and return to the **Profile Rib** dashboard.

(48) Note the magenta arrow indicating the direction of the rib connection. Click on the arrow to switch direction towards the boss. Enter **1.5** mm — the same thickness as the other ribs. The preview will identify whether the rib will connect to the boss (Figure 6.24).

(49) Click on the green tick icon (✔) to accept and close the feature.

(50) Save the model, i.e. **File > Save** (📄 Save).

6.9 Creating a Coaxial Hole Feature in the Boss

(51) Select the axis of the cylindrical boss in Figure 6.24 and then click on the **Hole** icon. The **Hole** dashboard opens.
(52) Press (CTRL + Hold) key and then click on the boss' top surface as an additional placement reference.
(53) Click on the drop-down depth menu (**Hole** dashboard) and select **Drill to selected...** option. Then pick the part internal base surface.
(54) Enter **5** mm hole diameter and close the **Hole** feature.

6.10 Creating Duplication of Features using Group and Pattern Commands

It would be very counterproductive to create the same feature several times in order to multiply it. As the reader has seen in the previous chapters (Chapters 4 and 5), this can be done more efficiently with the **Mirror** and even **Pattern** commands.

The **Pattern** (**Editing** group) can duplicate features in a rectangular array (Dimension pattern) or a circular array (Axis pattern). In both cases, the procedure expects a single feature to be selected and then patterned or mirrored using a reference datum for direction.

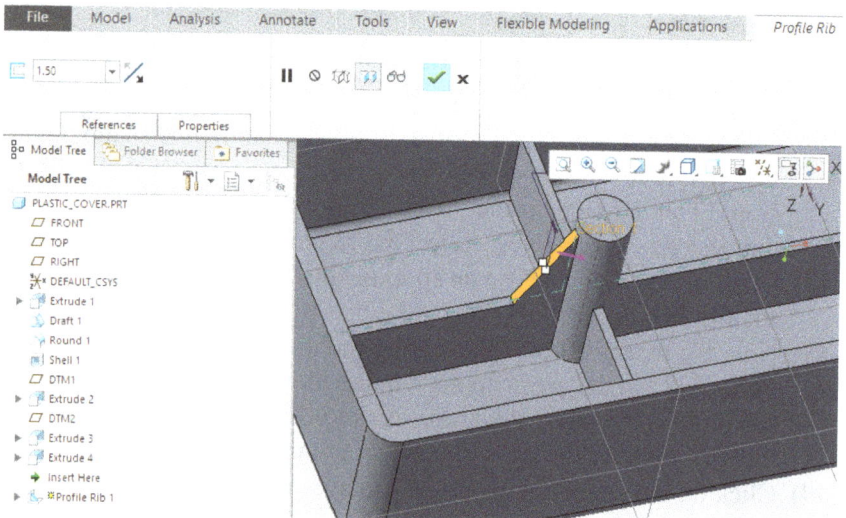

Fig. 6.24.   Profile Rib command dashboard.

The acceptable references for a **Pattern** command (feature) are as follows:

- Circular pattern — A datum (or feature) axis or cylindrical surface;
- Rectangular pattern — A single dimension for unidirectional pattern or two dimensions for a rectangular (two directional) pattern. Patterns can also be nested in features, or a pattern feature can be patterned itself.

6.10.1 *Patterning the triangular rib around the boss with a circular pattern*

(55) Select the **Profile Rib1** feature either from the Model Tree or from the Graphics area. Click on the **Pattern** ( ) icon, **Editing** group, to start. The **Pattern** dashboard will appear as shown in Figure 6.25.

(56) Click on the black down arrow in **Dimension** (top left corner of the dashboard) to pull down the menu, and select **Axis** (Figure 6.26). The black dots will indicate the locations of the pattern members.

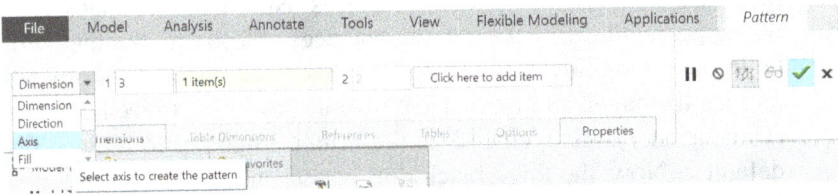

Fig. 6.25. Pattern dashboard.

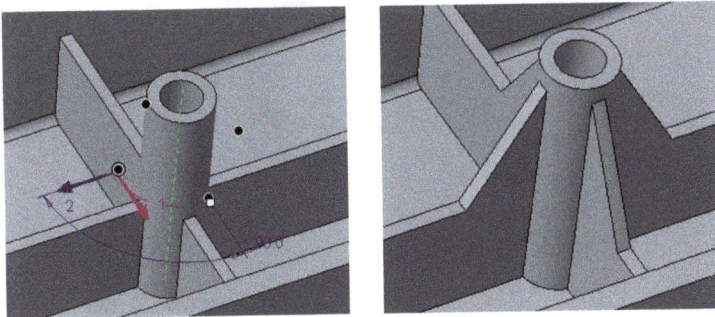

Fig. 6.26. Circular pattern of the triangular rib.

(57) Type in the number of pattern members and the angle between them in the input slots, i.e. **4** and **90** degrees.

(58) Click on the green tick (✔) to accept and close the feature.

6.10.2 *Creating a group of features and linear pattern*

(59) Press (CTRL + Hold) and click (LMB) on the features **DTM 2**, **Extrude 3**, **Extrude 4**, **Hole 1** and **Pattern 1** to select them from the Model Tree (Figure 6.27, left). The features to be grouped should form an uninterrupted sequence without any other features in between.

(60) With these features selected, click on the **Group** (🪣 **Group**) icon from the mini menu, or from the ribbon **Operations** group. A single **Group LOCAL_GROUP** feature will be created.

(61) With the **Group LOCAL_GROUP** feature selected, click on the **Pattern** (▦) icon, **Editing** group.

(62) The pattern dashboard will appear. Notice that the dimensions of all features in the local group will be shown (Figure 6.28). Click on the **30** mm dimension (datum plane offset) and type **30** mm to indicate the linear increment of pattern instances. Note that two black dots will appear, indicating the default instances of location, but they will be in the wrong direction. To correct this, click on **Dimensions** tab to open the panel, and type in **-30 (minus 30)** in the **Increment** slot (Dimension 1). Also, enter **3** as the instance number to replace the default **2**. Now, the three black dots appear in the correct direction (Figure 6.29). Notice that the **30** mm increment is equal to **DTM** 2 datum plane offset (rib location). The idea here is that the second

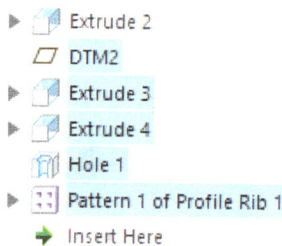

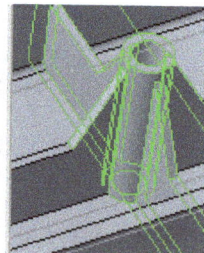

Fig. 6.27.   Selecting consequent features in the Model Tree.

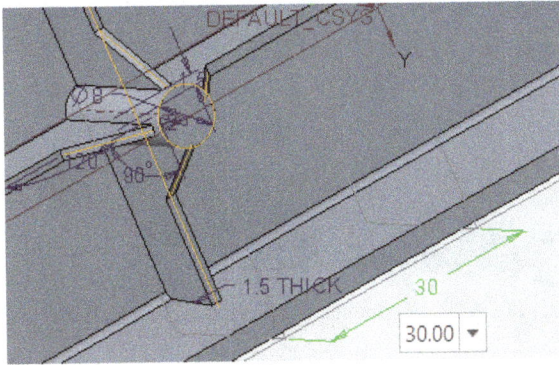

Fig. 6.28.   Linear pattern of a local group — increment input.

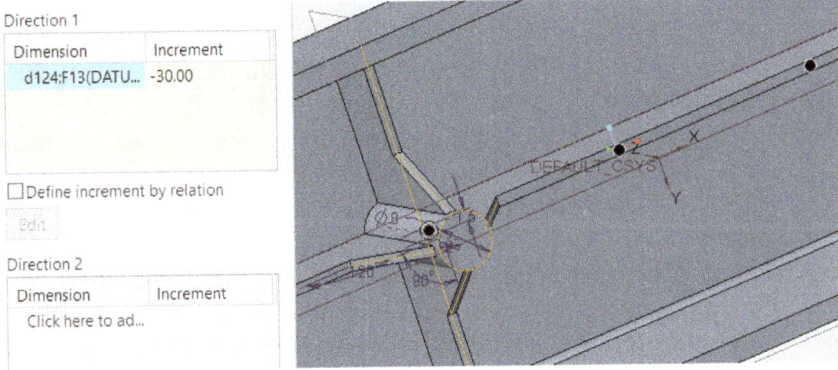

Fig. 6.29.   Linear pattern of a local group — preview.

instance will be in the middle, and the third will be symmetrical to the first. In general, the increment can be any value.

(63) Click on the green tick icon to accept and close. Three copies of all features in the group should appear in the model.

(64) Click on **File** > **Save** to save the current model.

6.11 Dynamic Modification of Model Dimensions

It is a routine procedure to modify dimensions and simultaneously view the resultant geometry of the 3D model. Creo has a very powerful engine that allows the user to dynamically modify a dimension, a sketch or something else and view the model changing in real time.

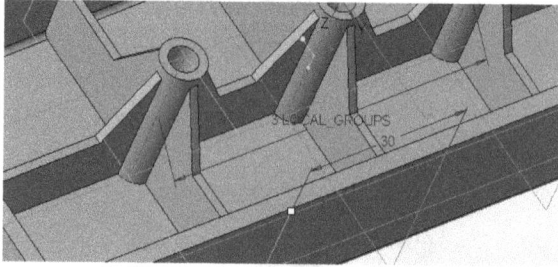

Fig. 6.30.   Dynamic modification of dimensions.

To demonstrate this capability let us try to move the ribs (plastic part) in new locations against the middle datum plane as follows:

(65) Select **DTM 2** and then click on the **Edit Dimensions** ($\overset{\longmapsto}{\mathsf{d1}}$) icon. The dimension **30** mm (datum plane offset) will appear. Click (LMB + Hold) on the dimension handle (white square) and then gently drag it to the left (Figure 6.30). Notice how the whole group of features will dynamically move following the Parent–Child links (Chapter 2).

(66) Double-click (LMB) on the **DTM 2** offset dimension and type in **40** mm. Now the ribs move to the left but have lost their symmetry.

(67) Now select **Pattern 2 of LOCAL_GROUP** from the Model Tree and click on **Edit Dimensions**. The number of instances **3** and their increment **30** mm will appear. Double-click on **30** to modify, and then type in **40**. Now the ribs will become symmetrical.

(68) Save the model, (**File > Save**).

6.12 Creating a Lip Using the Sweep Feature

Sweep is a feature that creates solid (or surface) 3D shapes by moving (sweeping) a 2D section along a closed or opened trajectory. Similar to **Extrude**, the **Sweep** requires a trajectory and a section. The **Sweep** trajectory can be a complex curve either located on a 2D datum plane (or flat surface) or a 3D curve.

The **Sweep** command can be activated from the model ribbon, **Model > Sweep** (**Shapes** group) or from the mini menu when a sketch feature is selected.

Continue with the PLASTIC_COVER.PRT development by creating a lip feature on the flange as follows:

(69) Click on the inner edge of the plastic cover flange. Note that only a single line section of the edge is selected (green colour). To select all sections from the loop, press the SHIFT button down and hold (SHIFT + Hold), then click (LMB) on another section of the inner edge. The entire edge becomes green in colour and can be selected as one.

(70) With the edge selected, click on the **Sweep** icon (Sweep). The **Sweep** dashboard opens as shown in Figure 6.31. The trajectory (edge) is highlighted in green. Two magenta arrows indicate the sketching plane orientation (normal to trajectory by default) and the **Sweep** start point.

(71) Click on the icon with a pencil ( ) (top left in the dashboard) to start (or edit) a sketch. The **Sketch** opens, and a cross of vertical and horizontal construction lines (dash dot, magenta colour) indicate the sketching plane and the **Sweep** start point.

(72) Click on the **References** icon (**Setup** group, top left corner) and then select the part's inner wall as a line reference (Figure 6.32, left).

The user can add sketch references by selecting any geometrical entities from the existing features, such as datum planes, axes, points, surfaces, edges, etc. The selected references are highlighted in blue and

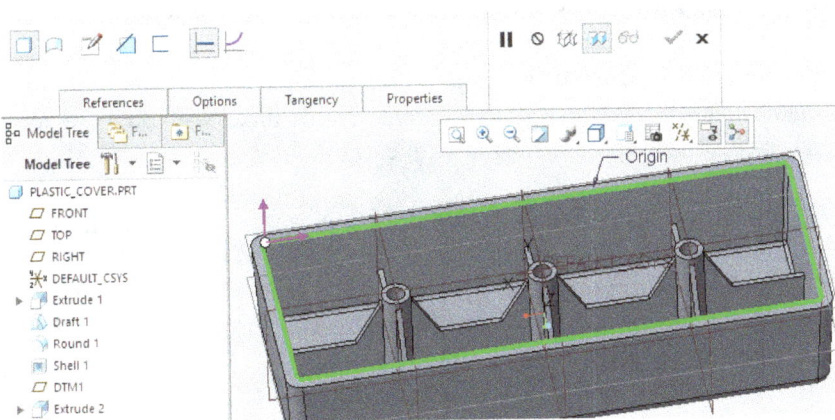

Fig. 6.31. Sweep dashboard (with trajectory selected).

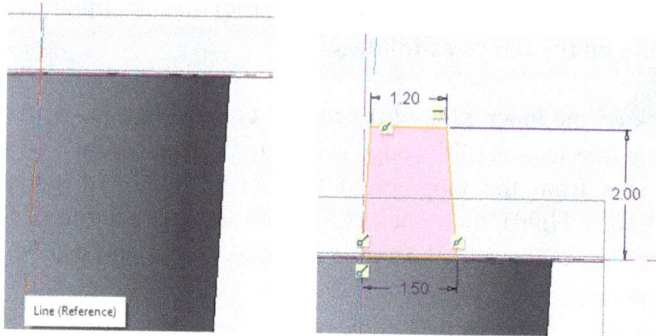

Fig. 6.32.   Sketching the sweep section.

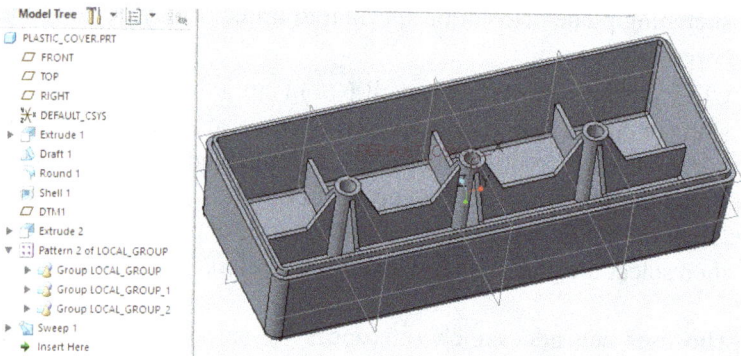

Fig. 6.33.   PLASTIC_COVER part and Model Tree.

appear in the **References** window. They work like 'anchors' and facilitate the location of the sketched entities. By using references, the user reduces the amount of dimensions and makes the modification process more predictable. The references create 'Parent–Child' links with the previous features that need to be managed with care.

(73) Pick the **Line Chain** tool and draw a closed section, starting from the cross point, following the line reference and the top of the cover as shown in Figure 6.32, right.

(74) Create dimensions and type in the values shown in Figure 6.32, right.

(75) Click on the green tick (✔) to close the Sketch. A preview of the feature appears in the dashboard indicating correct inputs.

(76) Click on the green tick (✔) to close the sweep feature. A view of the part and Model Tree after the **Sweep** are shown in Figure 6.33.

6.13 Applying Drafts to the Ribs and Bosses

All surfaces parallel to the pull direction should be drafted correctly with respect to the IM parting surface. The main walls of the model have been drafted already.

Continue the design by applying drafts of **1–2** degrees to both sides of all internal features such as ribs and bosses using the **Draft** command as described in Section 6.3.

(77) Press (CTRL + Hold) and click on the eight sides of the four ribs to select them as surfaces to be drafted. All should become green.

(78) Click on **Draft** icon to open the dashboard. Click on the **References** tab to open the panel (Figure 6.34). Note that the area under **Draft surfaces** has a message 'Individual Surfaces'. These are the pre-selected eight surfaces. Click within the pink slot under **Draft hinges.** The slot is now active and expects a selection of the <u>hinge</u> (a datum plane or a flat surface). Select the **TOP** datum plane as hinge (Figure 6.34). Note the current direction of the draft. The draft must be correct with respect to the IM parting plane. In order to see it clearly, increase the draft angle to **5–10** degrees and rotate the model to make sure that the draft is applied correctly. Click on **Reverse angle** ( ) from the dashboard or type in **-5 (minus 5)** to reverse the angle direction. If the result is correct, then type in the final angle, i.e. **1** degree.

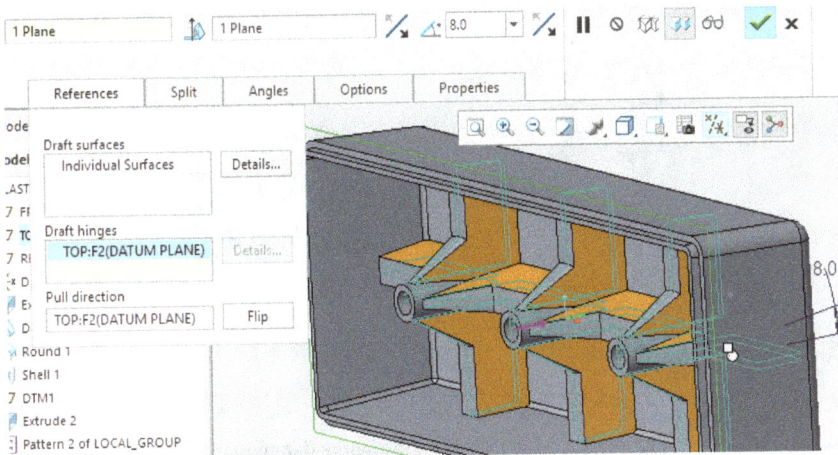

Fig. 6.34.   Drafts added to the ribs (Draft dashboard).

(79) Press the green tick (✔) to apply and complete the feature.
(80) Now add drafts to the three bosses. Select the cylindrical surfaces of all bosses. Click on the **Draft** icon and select the top face of the boss as the draft hinge. Note that in some previous Creo versions, the user has to have two semi-cylindrical surfaces for every single cylindrical surface (interpreted as two semi-cylindrical surfaces).

6.14 Create a Section View to Reveal the Internal Geometry

Before continuing in with the design, it would be helpful to view the model's internal geometry in order to analyse and measure some features and eventually correct accidental mistakes. One way to interrogate the model is by making sections or sectional views through the areas of interest. A section can be created through a datum plane or at an offset from a datum plane (or flat surface) as follows:

(81) Click on the **View** tab. Click on the **Planar** icon (▱) or alternatively click on the down arrow to expand **Section**, and select **Planar** option.
(82) Next, select a datum plane, i.e. **FRONT** to create a section. Drag the red arrow to offset the section and dynamically view the internal model details. Explore the **Section** dashboard options such as hatching, applying colour and other (Figure 6.35).
(83) Click on the green tick to close the **Section** tool. Notice that a new item **Sections** appears in the Model Tree under the **Insert Here**

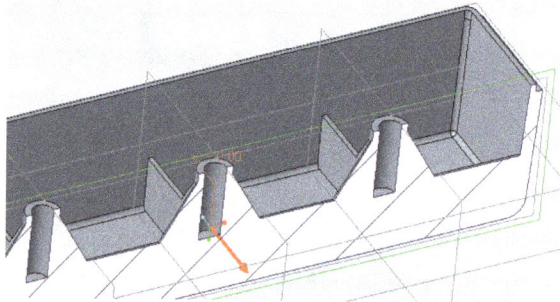

Fig. 6.35.   Model Section through a datum plane.

pointer. All sections that have been created are shown with their names. If a section is active, then the area of the model in front of the section will not be visible. Deactivate the section to view the whole model.

(84) Click on the section name to open the mini menu. The green square indicates an active section. Click it to deactivate (or activate). Use also (RMB + Hold) to redefine or delete a section.

(85) Save the model, (**File > Save**).

6.15 Additional Design Improvements

There are many considerations for efficient plastic design for IM, and this textbook does not intend to address all of them. However, some of the best practices are shown to help the reader in creating good design models.

One of the main rules is <u>uniform wall thickness</u>, which promotes a good and efficient design for plastic IM.

Creating internal <u>ribs</u> is an essential part of the design as they improve the part's structural strength and rigidity. The ribs are also used to connect the bosses to the walls and reinforce very tall internal details.

The <u>rounds</u> facilitate molten plastic flow and can add extra strength. However, the rounds can also thicken some areas and produce sink marks on the external surfaces. Therefore, the designer should be careful when applying rounds, especially rounds with a larger radius.

To avoid sink marks, the rib thickness at the connection with the walls should be at a maximum 50–60% of the shell thickness. The same rule applies to the bosses in the area of their connection with the part main shell.

Some additional features that can be created to improve the design with respect to IM process are as follows:

- Removal of extra material and maintaining constant thickness:
  In some designs, for instance similar to the Plastic cover, the holes in the bosses may not reach the bottom surface and there will be extra thickness at the base of each boss. This could create sink marks on the outside visible surface of the mounded part. A good remedy for this is either to extend the holes down to the part base surface or to create additional coaxial holes on the opposite side (under the boss) in order to reduce material concentration.

- Additional drafts:
  All surfaces that are parallel to the <u>pull direction</u> should be drafted in the correct direction. Surfaces that are under **1** mm may not have drafts.
- Adding rounds and chamfers:
  The rounds can improve the functionality, structural strength and overall appearance of the design. Also, rounds help to maintain a consistent thickness. In some cases, small rounds, for example with radius below **0.1** mm, are not needed because these will be created anyway during the tool manufacturing (cutting tools tip radius).
      Chamfers improve the part functionality and design for manufacture.

Figure 6.36 provides an example of how to improve the design following the above recommendations.

- Add a recycling symbol or other text on the part:
  It is a good idea to add text, images, company or other logos to the design that will be moulded on the plastic part at no extra cost. These features can be created using the tools available in Part modelling.

The next task (a challenge) will be to model a recycling symbol complete with material identification code so that each produced part will include this information for end-of-life recycling. You can source dimensions for standard recycling symbols from the Internet. An example is given in Figure 6.37, left image. The procedure is to draw the symbols on a sketch plane on the external floor surface of the Plastic_ cover. The sketch will be extruded above the surface by a small value of **0.3–0.5** mm.

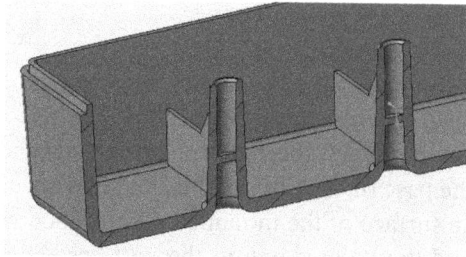

Fig. 6.36.   Additional features to improve Plastic cover design.

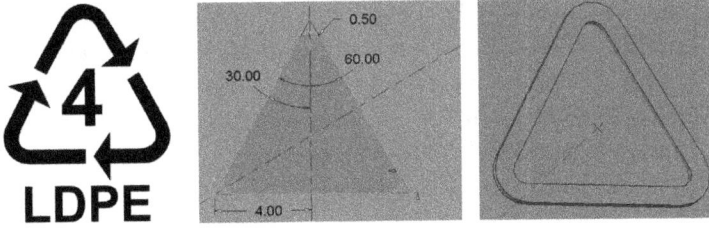

Fig. 6.37. Recycling symbol (left), sketch (middle) and the first extrude (right).

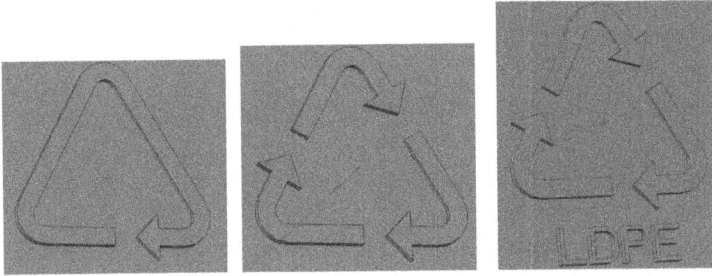

Fig. 6.38. Recycling symbol creation workflow.

Here is a possible workflow:

First, **Sketch** an equilateral triangle using three construction lines (click on the **Construction mode** to switch) and dimensions shown in Figure 6.37. Sketch also two **Centrelines** from the triangle vertices to the middle of the opposite side, and create a **Point** at their intersection. Switch back to normal line mode and sketch three lines (**Line Chain**) over the previous three construction lines. Round the corners with the **Fillet** tool (Figure 6.37, middle image). Use **equal** constraint on the three radii.

Create a feature **Axis** through the central point of the triangle sketch.

Next, use **Extrude** with **Thicken Sketch** option (**0.75** mm thick) and extrude the line by **0.35** mm in height (Figure 6.37, right image).

Use **Extrude (Remove material)** to create a slot interrupting the triangle and creating a space for the arrow. **Sketch** the arrowhead (at the end of the slot) and **Extrude** by **0.35** mm.

Group the last two extrudes into a single feature. Use **Pattern** to create a rotational pattern with three instances around the axis through the triangle centre. The last three steps are shown in Figure 6.38.

Another way to create the recycling symbol is to create a single arrow (with **Extrude** as above) and then use **Copy** and **Paste Special** (**Apply Move/Rotate transformations to copies** option) to multiply the extrude.

The sketch is quite complex, so you might also consider decomposing it into a series of simpler sketches/extrusions and rounds.

(86) Now save the model, (**File > Save**).

6.16 Analysis

Various types of analysis are available from the **Analysis** tab on the ribbon. For example:

- **Measure** — Provides useful tools to measure Distance, Length and Angle (2D), among others.
- **Model Report** — Offers tools for examining important characteristics such as Mass Properties (including Volume), Thickness, etc.
- **Inspect Geometry** (Tile) — Helps in analyses of drafts and other surface/curve properties.
- **Design Study** — Sensitivity, Feasibility and Optimisation can be performed on the model to find out how selected parameters from the model geometry will affect a defined criteria (mass, volume, stress, etc.)
- **Simulate Analysis** — Performs structural strength and thermal analysis using various materials, loads and constraints.

However, most of these analyses are outside the scope of this book, and so only some of these will be demonstrated in the lessons.

6.17 Model Check

In addition to the previous tools, the part design integrity can be checked as follows:

Start **File > Prepare > Model Check Interactive** to check and scrutinize the design geometry. A window will open in the Graphics area to show the results, as shown in Figure 6.39.

Hopefully, you will see only green and eventually yellow highlights. However, not all reported problems are detrimental, and the user has to use their judgement. Some of the warnings such as 'Short Edges' are more

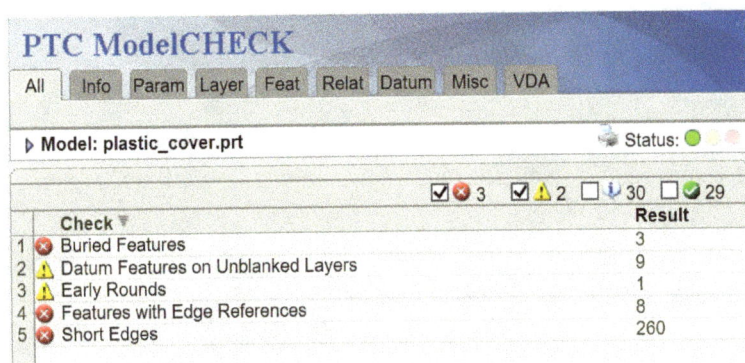

Fig. 6.39.   Model Check window.

informative than problematic. Typically, text features and small rounds are sources of short edges. Try to avoid them or keep them to a minimum.

However, the presence of some ***buried features*** can create problems and prevent mould generation, and therefore these must be properly examined. In this case, there are three buried features. After close examination, these are the three extrudes (Plastic_cover.prt) that have been drafted. The system flags them, but they are not a problem for the mould design.

To find the cause of a particular problem reported by the model check, click on it, and more options will be presented in the area below. Select the **Feature id** and then use **Highlight** to find it in the model structure. This may involve scrolling down the model check window.

Chapter 7

Assembly Modelling

7.1 Introduction

Almost every product design consists of several components that are connected together in a more complex structure called assembly. The assembly complexity can vary from a very simple two-part assembly, for example a bolt and a nut, to an assembly consisting of thousands of parts such as a car or a submarine assembly. Usually, assemblies that are more complex have a hierarchical structure of main parts and nested sub-assemblies logically connected together in a way that is relevant to the production. For example, a car's main assembly can be broken down to car body, engine sub-assembly, front suspension sub-assembly, seats, etc. Next, the engine sub-assembly can consist of engine block, piston sub-assembly, oil pump sub-assembly, etc.

Creo has a module called **Assembly** that allows multiple parts and/or sub-assemblies previously created to be connected together into an assembly model. In addition, the **Assembly** has advanced tools used to create different types of components directly in Assembly mode, such as parts, sub-assemblies, skeletons (special parts that are placed at the top of the assembly structure and used in Top-Down design to create the 3D layout of the parts within an assembly), etc.

A sub-assembly is in fact an assembly that is assembled into another assembly. Generally, every assembly can consist of parts only, parts and sub-assemblies, or sub-assemblies only.

Depending upon design complexity, the user can create a complex hierarchical tree of nested sub-assemblies and parts in order to express the design intent in the most logical and efficient way.

The links between parts and their locations are achieved by creating constraints or connections. These are defined as references between datum planes, coordinate systems, axes, points, surfaces, etc. One reference belongs to the assembly, while another reference belongs to the component that is being assembled (part or sub-assembly).

Similar to the part model, the assembly model has a list (model tree) of features (commands) that follows the design intent and creates the assembly structure with corresponding constraints or connections. The features and their references are arranged in a sequence and have specific Parent–Child links that ultimately affect the assembly model modification and subsequent regeneration.

The most common method to create an assembly is to assemble the components using a number of constraints. In this type of assembly, the parts are static and cannot simulate movement. This is called an assembly with constraints.

An assembly with movable parts (mechanism) has links between these parts that allows some degree of freedom. The links are called connections, and the assembly is called — Assembly with connections. An unconstrained part has 6 degrees of freedom: three translations and three rotations along the X, Y and Z-axes. Typical connection types that can be assigned in this assembly type are **Pin** (one degree of freedom allowing rotation), **Slider** (2 degrees of freedom for two translations), **Ball** (3 free rotations), etc.

In this lesson, the reader will learn how to create an assembly with constraints, reusing some of the parts created in the previous chapters.

Aim:
To develop modelling skills required to create assemblies of parts with correct constraints and predictable behaviour when modified and regenerated.

Outcomes:
At the end of this lesson, you should be able to:

• Create an assembly with constraints and understand how to apply various constraint types;

- Control the part orientation using the Dragger tool;
- Assemble parts using **Default, Coincident** and other constraint types;
- Modify parts in Assembly mode and regenerate the assembly;
- Understand Bottom-Up and Top-Down method of assembly;
- Analyse the assembly model for potential interference between parts;
- Create new parts in Assembly mode (Top-Down design).

7.2 The Assembly Model and Associativity Principle

The assembly model is associated with all parts and sub-assembly models that represent the components assembled within the assembly. As described in Chapter 2, Section 2.6, the assembly contains only information about their names, corresponding links and coordinates. The assembly file (*.ASM) does not have a record of the actual part geometry (features), which is the property of the *.PRT file. When an assembly is opened, the system looks for the file names of the components listed in the assembly Model Tree within the Working Directory. If a component is missing (or if it has been renamed), the assembly will fail to open and display the missing model geometry, and the name of the missing component will be shown in red colour in the Model Tree. To avoid this, it is good practice to locate all components of a project in a designated folder (directory) and to set it up as Working Directory. Another pitfall related to assembly modelling is renaming of some parts or subassemblies. The rule is never to rename any Creo object by means of Windows File Explorer tools.

7.2.1 *Renaming an assembly component*

An assembly component can be renamed when the assembly and the component are opened simultaneously in the same **Session**. The workflow is as follows: set up the Working Directory; **Open** the assembly; **Open** the component (part or assembly); click on **File > Manage File > Rename** and change the component name; **Save** the part; switch the Active window back to the assembly window and **Save** the assembly. Note that the assembly Model Tree will show the new component name.

After finishing work with one assembly project and before starting to work with another, always do the following: **Save** the assembly and all new components; **Close** all opened objects in the current **Session**; click on the **Erase Not Displayed** command to clean the memory; set up another Working Directory (for the second project) and then start the next project.

This way, the user will avoid mistakes and **Session** (memory) conflicts between objects with the same file name but different geometry.

7.3 Assembly Modelling Workflow

The assembly modelling workflow can be described as follows:

- Set up a Working Directory;
- Start a **New** assembly model (file);

Similar to <u>Part mode,</u> the <u>Assembly mode</u> is launched when a new model (assembly) is initiated (Figure 7.1). The software will open the assembly modelling window and ribbon interface with tools for creating an assembly (Figure 7.2).

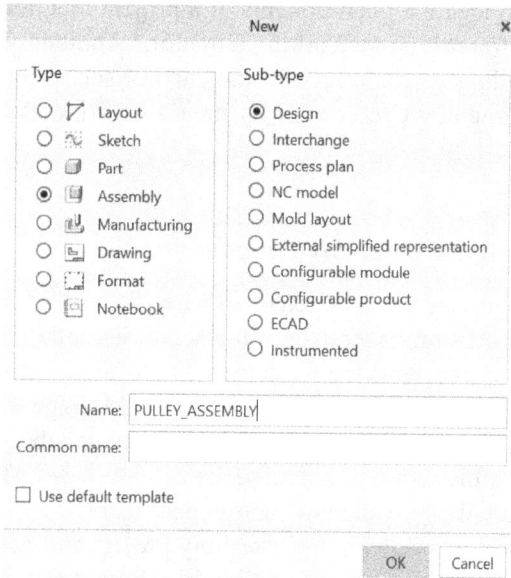

Fig. 7.1.   New assembly window.

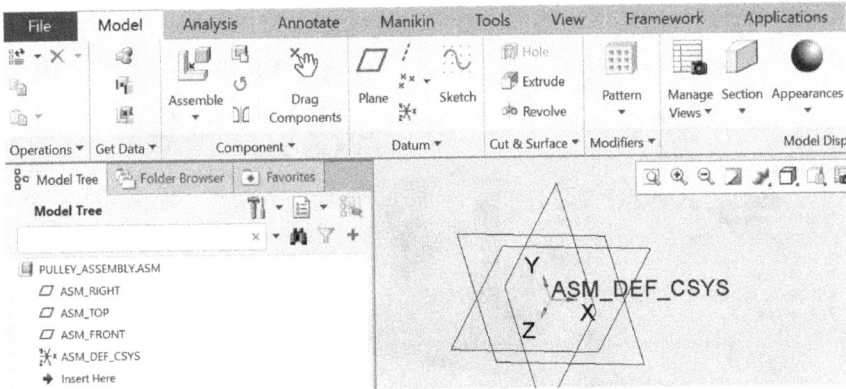

Fig. 7.2.   The assembly modelling window.

- Create extra datums (datum planes, axes, points, coordinate systems, etc.) and features if they are needed as additional references for components that will be assembled;
- Assemble the first component;
- Assemble the next component;

A component (part or assembly) is assembled using the **Assemble** command (**Model > Assemble**) from the assembly ribbon interface (Figure 7.2). The Working Directory opens and the user can select a component to be assembled. After that, the **Component Placement** dashboard appears, which allows the user to define the component location and relationship with the assembly. The dashboard has a variety of tools for applying placement constraints to the selected component in relation to the assembly (Figure 7.3). Applying constraints is an essential step in the assembly process and will be discussed in the next section.

7.3.1 *Constraints theory*

A constraint defines how a part is located within an assembly and specifies its relative position by means of a pair of references. With the **Component Placement** dashboard opened, one reference should be selected from the assembled part and the second reference either from the assembly of from another assembled component. The constraints are applied one at a time until the component is fully constrained. For each

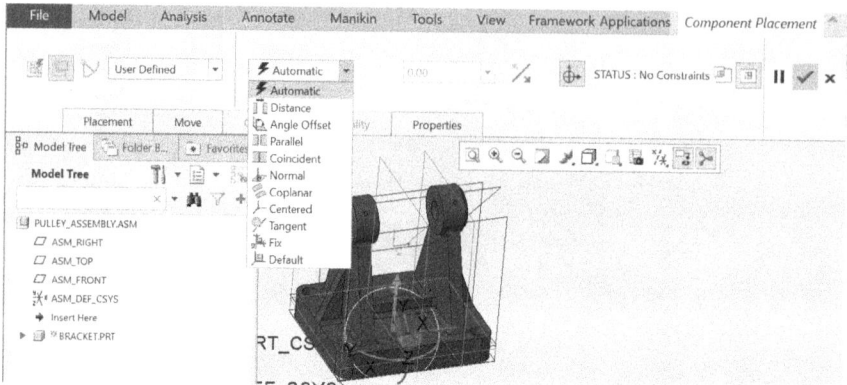

Fig. 7.3.   Component Placement dashboard.

type of constraint, the applicable references can be planar or rotational surfaces, coordinate systems, datum planes, axes, straight edges, points, etc. Usually, several combinations of constraints are needed to define the component placement and achieve 'Fully Constrained' message in the **STATUS** dashboard slot. The main principle in this process is to restrict all possible component movements. In other words, all six degrees of freedom (three translations and three rotations) should be constrained. For instance, this can be achieved by selecting a pair of coordinate systems — one belonging to the part and another to the assembly — or by selecting three pairs of orthogonal datum planes (or planar surfaces) to be coincident, parallel, normal or at a distance from one another.

The available constraints in the **Component Placement** dashboard are as follows (see Figures 7.3 and 7.4):

- **Automatic** — This tool will try to capture the design intent and choose the most relevant constraint type depending on the selected references. For example, if a pair of planar references (datum planes or planar surfaces) are selected, then **Coincident** or **Distance** constraint will appear in the **Placement** tab list.
- **Distance** — This constraint applies a parallel condition with a distance between the two selected references. The references can be pairs of planar surfaces, datum planes, axes or combinations of these. The distance between the references is displayed with a <u>handle</u> (small white colour square) that can be dragged to translate the component. The distance can be changed directly from the **Offset** slot in the

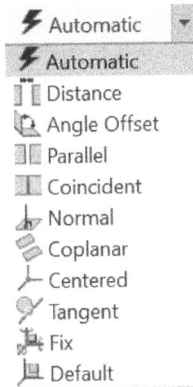

Fig. 7.4. Constraint types menu.

Placement panel of the dashboard. Click on the **Placement** icon to open the panel. Also, the component orientation can be changed with the **Flip** icon.

- **Offset Angle** — This is an angular constraint between a pair of intersecting planar surfaces, datum planes, axes or combinations of these. The angle between the references is displayed with a handle that can be dragged to rotate the component. The angle can be entered directly in the **Offset** slot, in the **Placement** panel.

- **Parallel** — Similar to **Distance,** this constraint imposes a parallel condition between two selected references. However, the distance between the references is not fixed and can vary.

- **Coincident** — The selected pair of planar surfaces, datum planes, or two axes will be coincident.

- **Normal** — The selected planar surfaces, datum planes, axes or combinations of these will be perpendicular (normal) to each other.

- **Coplanar** — This constraint positions a selected edge, axis or surface on a selected planar reference.

- **Centred** — This constraint is used to align the axes of revolved components, revolved surfaces, or holes. The assembled component will keep its orientation and will share the same axis.

- **Tangent** — It imposes a tangency between selected circular surfaces.

- **Fix** — It fixes the current position of the assembled component.

- **Default** — It aligns the default coordinate systems of the component and assembly.

As noted previously, it is desirable to achieve a 'Fully Constrained' status in the **Component Placement** dashboard before completing the feature. However, it is possible to close the dashboard even though the component is not fully constrained, with a status 'No Constraints' or 'Partially Constrained'.

7.3.2 *3D dragger — tool for initial component position*

The 3D Dragger (Figure 7.5) is a tool that can be used to translate or rotate the assembled component along the three main axes X, Y and Z. It does appear immediately after the opening of the component. The 3D Dragger is attached to the part and has three arrow handles for translation and three circular handles for part rotation. The user can position the cursor over a handle, pick it with the (LMB + Hold) and drag the handle in order to perform the corresponding movement.

Use this tool to provide an approximate initial part position. The constraints will take it into account and provide a relevant placement.

Often, it is difficult to pick the correct references when the assembled component datums overlap with the assembly datums. In such cases, use the Dragger to move the component away in order to separate the datums and select the correct reference. Alternatively, use **Pick From List**.

When a constraint is set, it disables the corresponding 3D Dragger movements. Use the 3D Dragger to test which movement is still possible

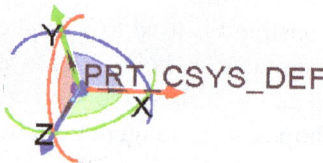

Fig. 7.5.   3D Dragger.

and apply the corresponding constraint type until 'Fully Constrained' status is achieved.

7.4 Bottom-Up Assembly Method

The Assembly mode has tools that enable several assembly methods. The most common and simplest method is called the Bottom-Up assembly design. In this approach, the parts should be completed in advance using the Part mode and then assembled. The result is a simple, straightforward and predictable-for-modifications assembly. However, a major disadvantage is that the component design with all dimensions must be available to construct the features, which is often impossible in a new project. In addition, the modifications of common dimensions in two or more parts within one assembly will be time consuming and prone to errors. This reason for this is that for one common dimension the user should modify at least two parts. For instance, in the BRACKET and SPINDLE assembly, the hole diameter in the BRACKET should match the SPINDLE diameter. In order to modify the common diameter, the user must open and modify both parts separately. This could be more difficult if there are many common dimensions within an assembly. The Top-Down assembly method, illustrated later in the book, can help solve this problem.

7.4.1 *Creating pulley assembly using default and coincident constraints*

In the next sections, the Bottom-Up assembly method will be demonstrated with the three parts SPINDLE.PRT, BRACKET.PRT and PULLEY PRT, which were shown in Figure 4.1. They should have been created following the lessons in the previous chapters.

(1) Start Creo (unless it is already running).
Set up Working Directory. Click on **File > Select Working Directory > C:\USER\CREO_PRACTICE**.

(2) Start a new Assembly. Click on **File > New**, or click the **New** ( ) icon from the Quick Access Toolbar. Select **Assembly** as **Type** in the **New** window (Figure 7.1) to launch the Assembly mode.

(3) Type PULLEY_ASSEMBLY in the **Name** slot. Remove the tick mark for **Use default template** (unless it is clear what is the default template) and then select the **OK** icon (OK) to accept.

(4) The **New File Options** window opens, where a template with the appropriate System of Units can be selected. Use **mmns_asm_ design** for Metric (metre, Newton, second) and click **OK** (OK).

The assembly modelling window and ribbon with tools opens. The Model Tree contains the assembly name only. To reveal the features, go to **Settings** (🗂) and select **Tree Filters** to invoke the **Model Tree Items** window. Select **Features** (drop-down menu in the Model Tree), and then click on **OK**. The three default assembly datum planes ASM_RIGHT, ASM_TOP and ASM_FRONT and the coordinate system ASM_DEF_ CSYS will be revealed (Figure 7.2). These datum features will be used as initial assembly references.

(5) Click on the **Assemble** icon (🗏) in the **Component** group, i.e. **Model > Assemble**.

(6) In the **Open** window, select BRACKET.PRT and click **Open**. The **Component Placement** dashboard appears. Notice that the **STATUS** slot indicates 'No Constraints' (Figure 7.3).

(7) Click the drop-down menu in **Automatic** to reveal all available constraint types (Figure 7.3).

(8) The very first component in any assembly can be placed directly using the **Default** constraint. Note the 'Fully Constrained' message.

📝 The **Default** constraint places the component at the assembly origin, aligning their internal (system created) coordinate systems. This constraint does not use any references, and there is no Parent–Child relationship.

(9) Click on the green tick icon (✔) to apply the changes and close the dashboard. The component name is now shown in the Model Tree.

ⓘ Notice:

• All constraint types described in Section 7.3.1 are available from the dashboard drop-down menu, shown in Figures 7.3 and 7.4. Instead of

Automatic, the user can directly select/apply a relevant constraint type.

- Click on the **Placement** tab in the **Component Placement** dashboard to open the panel with a list of applied constraints and corresponding references. To edit, select a constraint and press (RMB + Hold) to open the menu with **Delete** or **Disable** options.
- Click on **New Constraint** to start a new constraint set.
- Click on a constraint reference and press (RMB + Hold) to open the menu with **Remove** or **Information** options.
- From the **Placement** window, you can switch from one constraint to another and change the combination of applied conditions. For example, from **Coincident** to **Distance**, from **Normal** to **Parallel**, etc.

It is possible to close the **Component Placement** dashboard even with 'No Constraints' or 'Partially Constrained' message. The Model Tree will record the assembled component. To redefine and correct its placement, select the component, and from the mini menu select the **Edit Definition** ( ) command. Another way to invoke the **Edit Definition** command is to select the component and then press (RMB + Hold).

Coincident constraint enables a component to be assembled by:

(a) Coinciding the axes of cylindrical or conical surfaces belonging to the component with another that belongs to the assembly. Select axes or cylindrical (conical) surfaces as references.
(b) Coinciding datum planes or planar surfaces selected from the component and assembly.

(10) The second part will be assembled using a **Coincident** constraint.

Click on the **Assemble** ( ) icon from the ribbon. In the **Open** window, select SPINDLE.PRT and click on **Open**. The **Component Placement** dashboard appears. Notice the 'No Constraints' status.

(11) Click on the black arrow next to **Automatic** to open the constraint drop-down menu and select **Coincident**.
(12) Move the mouse cursor above the component and notice that the relevant references start flashing, i.e. being pre-selected. Click on the SPINDLE.PRT cylindrical surface. If selected correctly, the

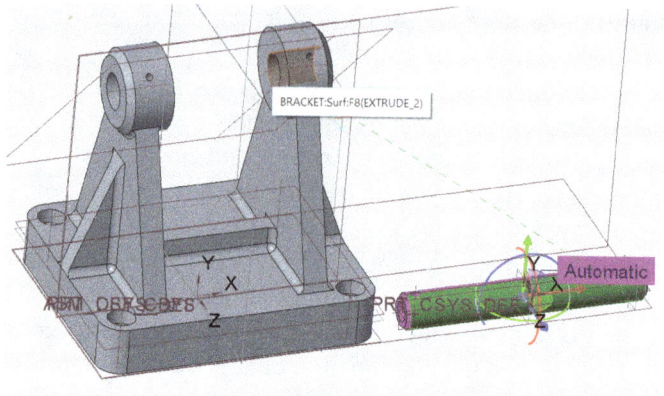

Fig. 7.6.   Selecting references (in green colour).

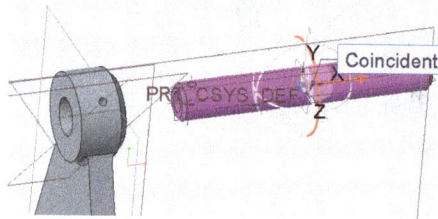

Fig. 7.7.   SPINDLE.PRT with Distance constraint.

surface will become green (Figure 7.6). Notice that a thin dashed line will indicate a link to the second selection.

(13) Next, move the cursor to one of the BRACKET support holes and click on it to select. If correctly selected, the spindle axis will become coincident with the hole axis, as shown in Figure 7.7.

(14) Select the **Placement** tab in the dashboard to reveal the references (Figure 7.8). If correctly selected, one reference should belong to the component (SPINDLE) and another to the assembly or to an assembled component (i.e. BRACKET). Notice that the status is 'Partially Constrained', indicating insufficient constraint sets.

(15) Move the mouse to the SPINDLE part and select the middle vertical datum plane. Next, select either the ASM_RIGHT or RIGHT (BRACKET) middle datum plane. The status is 'Fully Constrained', but the SPINDLE is offset and not properly assembled.

Fig. 7.8.   Placement tab and Coincident constraint references (SPINDLE).

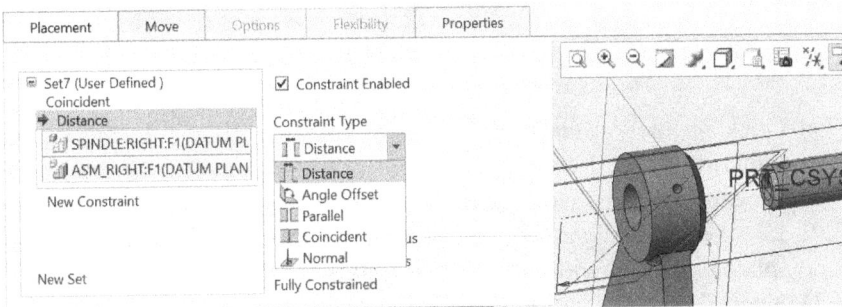

Fig. 7.9.   Constraint Type (in Placement tab) and alternative constraints.

(16) Click on the **Placement** tab and notice that the second constraint has been automatically set to **Distance**, assuming that the component reference is offset from the assembly reference by a distance, as shown in Figure 7.9. To correct, click on the drop-down menu under **Constraint Type** and switch from **Distance** to **Coincident** (Figure 7.9). The component now assumes the correct position.

(17) Click on the green tick icon (✔) to apply the changes and close the dashboard. The assembly should look as shown in Figure 7.10.

When moving the mouse cursor over part geometry (surfaces, axes, planes, etc.) to pre-select an assembly reference only the relevant (to the specific constraint type) entities will flash and be available for selection.

(18) The next component to assemble is the PULLEY.PRT. It will be placed using the same method as in SPINDLE.PRT.

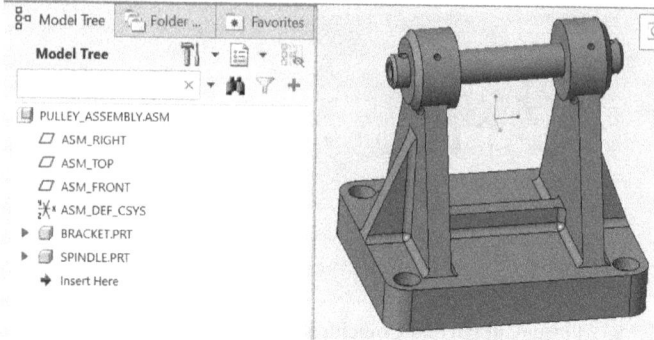

Fig. 7.10.   PULLEY_ASSEMBLY.ASM with SPINDLE.PRT assembled.

(19) Click on the **Assemble** ( ) icon from the ribbon. In the **Open** window, select PULLEY.PRT and click **Open**. The **Component Placement** dashboard opens. Notice the 'No Constraints' status.

(20) Use the 3D Dragger (Figure 7.5) to move the part closer and approximately aligned with the SPINDLE axis.

(21) Keep the constraint selector in **Automatic** and move the mouse cursor above the component feature to preselect, i.e. until the PULLEY central hole flashes, and click (LMC) on it to select. The surface will turn green in colour.

 Zoom-in the area around the central hole if the mouse cursor cannot select the surface properly.

(22) Move the cursor to the BRACKET external support cylindrical surface and click on it to select the second reference. If correctly selected, the surface will turn green in colour for an instant and then the pulley will move and become coaxial with the bracket hole.

(23) Click on the **Placement** tab (dashboard) to reveal the selected references (Figure 7.11). If correctly selected, one reference should belong to the component, i.e. PULLEY, and the second should belong to the BRACKET. Notice than the **Constraint Type** is set to **Coincident**. If this is not the case, then click on the black arrow under the **Constraint Type** to open the drop-down menu and select **Coincident**. The status is 'Partially Constrained', indicating insufficient number of constraint sets.

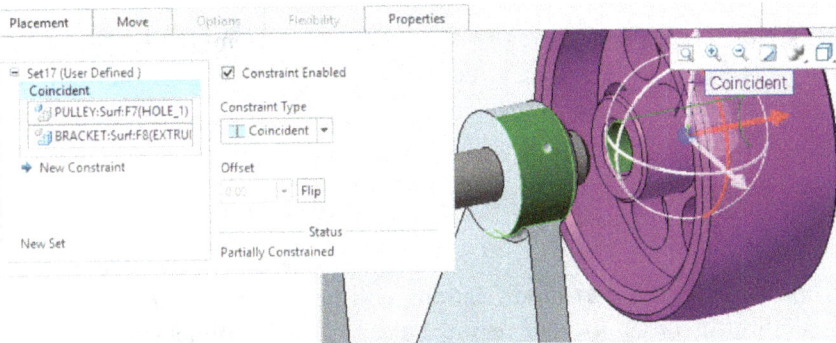

Fig. 7.11. New Constraint in the Placement tab (PULLEY placement).

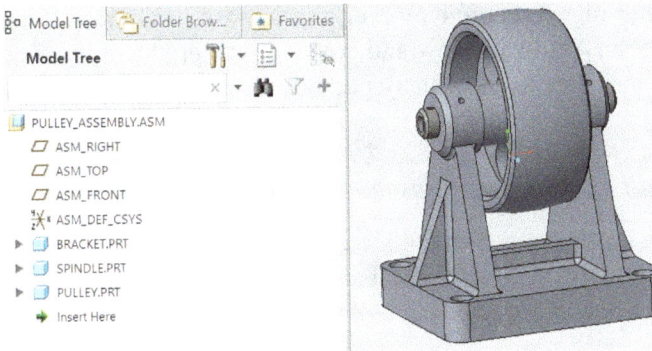

Fig. 7.12. PULLEY_ASSEMBLY.ASM with PULLEY.PRT assembled.

(24) To complete, click on the **New Constraint** (New Constraint) at the bottom of the **Placement** tab and open a new constraint set. The **Constraint Type** is set to **Automatic** by default.

(25) Move the cursor to the PULLEY part and select the middle vertical datum plane (RIGHT) and then select the BRACKET middle datum plane. The status is now 'Fully Constrained'; however, the PULLEY is offset from its correct position.

(26) To correct, click on **Placement** tab, and change **Constraint Type** from **Distance** to **Coincident**. Now the component should assume its correct position.

(27) Click on the green tick icon (✓) to close the dashboard. The assembly should look as shown in Figure 7.12.

(28) Now click **File > Save** (![Save icon] Save) to save the assembly file. Notice the 'PULLEY_ASSEMBLY has been saved'. message in the message area at the bottom of the screen.

7.5 Part Modifications and Global Interference Analysis

Almost inevitably, part modifications will be required after creating the assembly. In order to change the values of some dimensions (feature parameters), note that every part of an assembly can be opened individually in Part mode and modified accordingly. Following the associativity principle (Chapter 2), the assembly will assume the changes and update after regeneration. However, this approach will lead to frequently switching the Active window and could reduce productivity, especially if there are modifications on many parts of the assembly.

7.5.1 *Edit dimensions in assembly mode*

There is a quicker way to modify the parts in Assembly mode. In addition, it would be useful to observe immediately how the changes would affect the whole assembly. This approach will be demonstrated in the next example.

(1) **Start** Creo (unless it is running).
(2) Set up the Working Directory (pulley assembly and part files location).
(3) Open the pulley assembly model created previously, i.e. **File > Open** and select PULLEY_ASSEMBLY.ASM.
(4) Expand the Model Tree to reveal part and assembly features. From the Model Tree settings (![icon]), select **Tree Filters** to invoke the **Model Tree Items** window. Click on **Features**, and then **OK**.
(5) Click on the **Datum Display Filters** (![icon]), Graphics toolbar, and un-select all in order to hide the datum display.
(6) In the Model Tree, click on the black arrow (▶) in front of the PULLEY.PRT to reveal the list of part features (the model history).

(7) Now find the **Revolve 1** feature and click on the black arrow (▶) in front of it. This will reveal the **Sketch 1** that is embedded in it.

(8) Click on **Sketch 1** and then select **Edit Dimensions**. All dimensions related to this sketch will be displayed.

It is possible to perform the same modification directly from the external **Sketch 1** because it is the same sketch as the one in **Revolve 1**.

(9) Find the **50** mm horizontal dimension (the pulley wheel width) and double-click on it to open the modification dialogue slot. Notice that the ribbon will change to **Dimension** interface providing advanced control on dimension value, tolerances and dimension display. Type in **60** and press the ENTER key to confirm.

(10) Go back to the **Model** ribbon and click **Regenerate** icon ( ) from the Quick Access Toolbar in order to update the whole assembly.

7.5.2 *Interference analysis*

Due to the previous modification, the PULLEY width is now larger. However, there is no apparent change in the assembly appearance. The next steps show how to perform interference analysis on the assembly.

(11) Click on the **Analysis** tab from the assembly ribbon and select **Global Interference** (**Analysis > Global Interference**). This tool measures a possible interference between any parts. In the **Global Interference** dialogue window, click on **Preview** icon. The analysis will display that **Part 1** (BRACKET) has a common volume of **5160** mm^3 with **Part 2** (PULLEY). The interfering areas are displayed in red colour in the Graphics area (Figure 7.13).

(12) Click **OK** to close the **Global Interference** window and return to the **Analysis** interface.

A 3D C AD assembly with interference is an unacceptable design. If the pulley assembly geometry or drawings are exported and the parts are manufactured with their current dimensions, they may not fit properly in the real assembly. Therefore, all parts that interfere should be analysed and corrected accordingly.

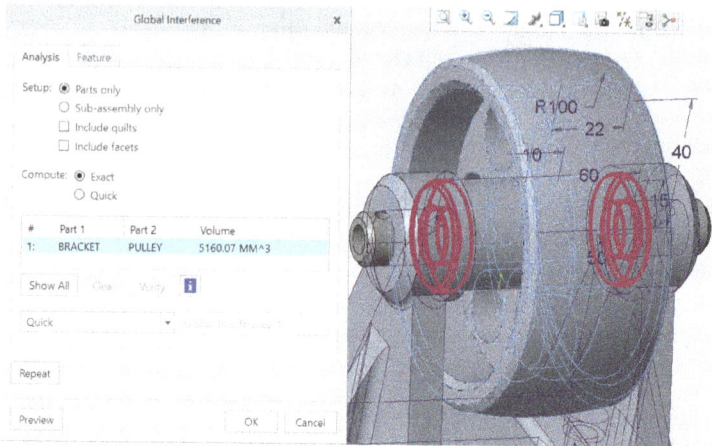

Fig. 7.13.   Global Interference analysis.

7.5.3 *Measuring dimensions*

To correct the interference in the previous scenario, the user could either increase the width between BRACKET supports or reduce the PULLEY width. In order to change the BRACKET, it would be useful to measure the exact distance between supports.

(13) Use **Hide** and **Show** commands to hide and reveal back a selected component from the Graphics area and facilitate entities selection. To hide a part, go to the Model Tree and select PULLEY.PRT and then click the **Hide** icon (✎ Hide) from the mini menu. The part will be hidden. To display it back, select the part name and then click the **Show** icon (◉ Show).

(14) From the **Analysis** ribbon, go to **Measure** and click on the down arrow (▼) to reveal the following measuring tools: **Distance**, **Length**, **Angle**, **Diameter**, etc. Each tool requires corresponding references.

(15) Click on **Distance**. This tool expects two references and measures the distance between them. Move the cursor to the inner circular support surface and click on it to select. The surface will assume a green colour. Press (CTRL + Hold) and then select the opposite surface. The distance between these surfaces displays on the screen (Figure 7.14).

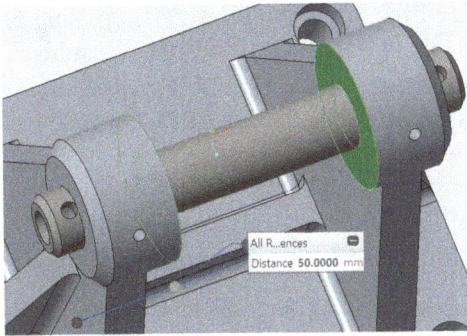

Fig. 7.14.   Measuring distance.

(16) Select the PULLEY.PRT from the Model Tree and then click on **Show** (the mini menu) to view the part. Select **Sketch 1** and modify the width dimension to **45** mm. **Regenerate** the model.

(17) Perform **Global Interference** analysis again to make sure that there are no interfering parts.

(18) Save the assembly file, **File > Save** (💾 Save).

💡 Saving the assembly model will also save all parts that have been modified in the current **Session**.

7.6 Creating Components in Assembly Mode (<u>Top-Down</u> Method)

ℹ️ <u>Top-Down</u> method in assembly allows the CAD designer to control the assembled parts from a top level by means of a variety of tools such as skeleton parts, assembly sketches, datum planes, and other features positioned at the top of the assembly Model Tree and used as references for creating new parts or features. Because of the Parent–Child relationships, the modifications at the top will drive modifications at lower levels though common references, hence the name <u>Top-Down</u> approach. The advantage is that the designer can control a complex assembly from one place at the top and achieve a well-structured and predictable-to-modifications design. A disadvantage is that this approach has less flexibility and requires good understanding of the final design and relationships between parts. One variant of the Top-Down design is creating new parts directly in Assembly mode using references from the assembly.

Suppose that the previous assembly needs two plastic washers, located at each side of the PULLEY.PRT, to reduce friction. The next steps describe how two new parts will be created in Assembly mode.

(19) Set up a Working Directory and then **Open** the PULLEY.ASM.

(20) Change the Model Tree settings to reveal component features.

(21) From the **Component** group, select **Create** ( ) icon to create a new component in the assembly. **Create Component** window will open. Keep the default selection and type the part name WASHER as shown in Figure 7.15, left. Click on **OK** to continue.

(22) In in the next **Create Options** window (Figure 7.15, right), select **Empty** (or **Copy from Existing**) and click on **OK**. The new part (WASHER) will appear in the Model Tree.

 Note that the new part will assume the same System of Units as the current assembly, i.e. millimetres (**mmns_part_solid** template).

(23) Select the WASHER in the Model Tree. From the mini menu that will appear next to the part, click on the green knot (**Activate**) to **Activate** the part. All inactive parts are dimmed in the Graphics area.

 The new part created in Assembly mode does not have any features, and they need to be developed. By activating the part, Creo switches to

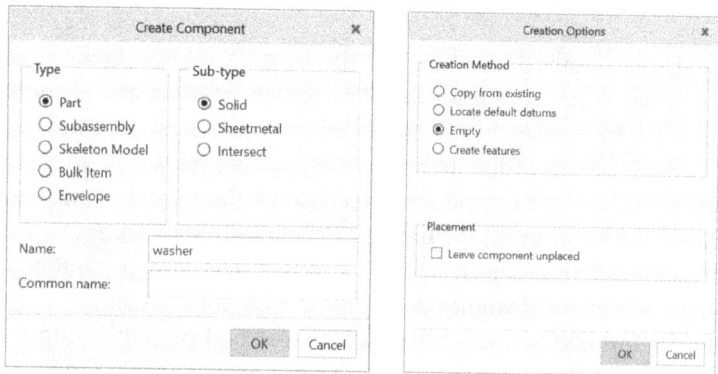

Fig. 7.15.   Create component in Assembly mode.

<u>Part mode</u> and the designer can use all commands and tools available to create the part geometry. The inactive parts will be dimmed; however, their geometry and datum features are available to be selected as references for the active part. The benefit of this approach is that the geometry and datums of previously assembled components can be selected as references for the new part features. Thus the dimensions of the new part will be linked with the existing features and therefore their modifications will be automatically driven by the previously created features.

<u>The main rule in this method</u> is that all references should be picked only from those that are located <u>above the active part</u> in the Model Tree, and never from <u>below the current active part</u> (if there are other components after the active part). Often, the needed references coincide with one another, and so the user can pick a wrong reference by mistake from <u>under the active part</u>. This will create a <u>circular reference</u>. It means that a <u>Parent</u> feature will inherit a <u>Child</u> feature reference. This is dangerous because a <u>circular reference</u> might have unpredictable effects on the part geometry after modification and regeneration of the whole assembly. If this happens, the system will immediately create a report file (*.CRC) in the current Working Directory to indicate all <u>circular reference</u> loops. A yellow warning will appear at the bottom of the Graphics area after regeneration. The user must revisit and correct all affected features, which is a very difficult task, even for an experienced user. It is much easier to avoid circular references by checking the position of the references and corresponding part against the active part. Before selecting a reference, always press (RMB + Hold) and select the right reference from **Pick From List** option.

(24) Select BRACKET.PRT from the Model Tree and then click **Hide**.
(25) Select the left-hand side surface of the PULLEY.PRT (Figure 7.16) as the sketching plane and click on **Extrude** icon (from the mini menu) to start the feature. The **Sketch** interface with drawing tools opens.
(26) Select **Concentric** tool, first click on the SPINDLE cylindrical surface as reference (for the circle centre) and then sketch two concentric circles. Avoid snapping the circle radius to any other references. Type in **15.5** mm and **28** mm as diameters for the inner and outer circle. Notice that horizontal and vertical references are not required as the circle centres are defined by the SPINDLE cylindrical surface.

CAD/CAM with Creo Parametric

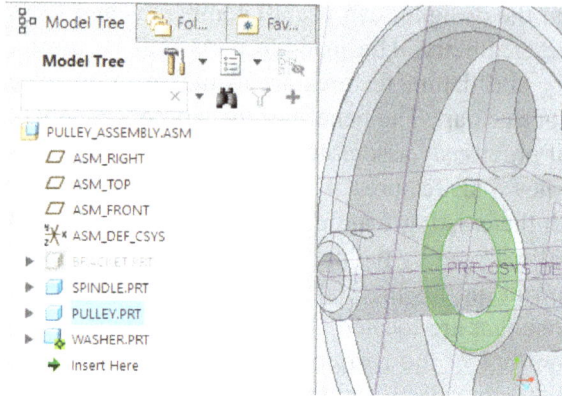

Fig. 7.16.   Selection of the sketching plane for the Extrude (WASHER.PRT).

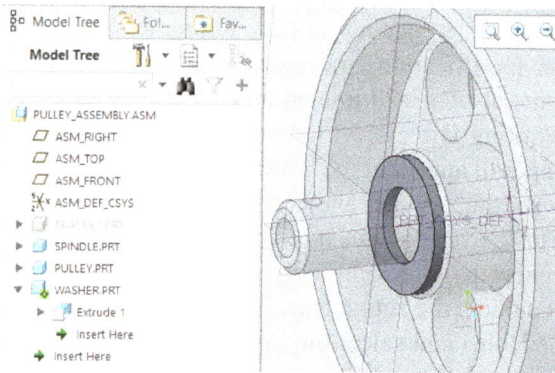

Fig. 7.17.   The WASHER.PRT created in Assembly mode.

(27) Click on the green tick (✔) to leave the Sketcher.

(28) Type in **2.5** mm as depth in the **Extrude** dashboard and then click the green tick (✔) to close the feature.

The WASHER.PRT is shown in Figure 7.17.

(29) To return back to Assembly mode, select the assembly in the Model Tree and click on the green knot from the mini menu to **Activate**.

(30) Save the assembly, **File > Save** (⊞ Save). The **Save** command will save the assembly file, all modified and all new parts created in Assembly mode, in the Working Directory.

7.7 Mirror Component Command — Duplicating Parts in Assembly Mode

In order to add a second WASHER.PRT to the assembly, the user can assemble the same part again on the other side of the pulley using constraint and following the assembly procedure given in Section 7.4.

However, it is quicker to duplicate the WASHER.PRT with the **Mirror Component** command in <u>Assembly Mode</u> as follows:

(31) Select the WASHER.PRT and click on the **Mirror Component** icon (⬛) from the **Component** group.

(32) The **Mirror Component** dialogue window will appear. Switch the selector from **Create a new model** to **Reuse selected model**. Pick the **Mirror plane** slot to activate it (and make it ready for mirror plane selection). Move the mouse cursor to the Graphics area and select ASM_RIGHT datum plane. Click on **Preview** box to preview. If the result is satisfactory, click **OK** to close the command.

(33) Select WASHER.PRT and click on **Open** from the mini menu to open it in a new window.

(34) Select the washer name (at the top of Model Tree), then go to **View** > **Appearances** and select blue colour (or another colour).

(35) Switch the <u>Active window</u> back to PULLEY_ASSEMBLY.ASM, select BRACKET.PRT from the Model Tree and click on **Show**.

(36) Save the assembly, i.e. **File > Save**.

7.8 Modifying the Pulley Assembly and Adding More Parts

Next, we are going to modify the SPINDLE.PRT in order to create space for two additional steel washers at each end and two split pins (cotter pins).

(37) Open PULLEY_ASSEMBLY.ASM (unless it is already opened).

(38) Reveal the features, expand SPINDLE.PRT, select **Extrude 1** and click on **Edit Dimensions** (from the mini menu) to show the spindle length. Double-click on the current value (**105** mm) and

enter new value of **108** mm to accommodate two steel **1.5** mm washers.

(39) Click **Regenerate** icon ( ) to update the assembly. The spindle length will increase by **3** mm. However, the holes at both spindle ends have not moved to provide room for the washers.

(40) Now select **DTM 1** datum plane, SPINDLE.PRT, and click on **Edit Dimensions**. Double-click on the current value (**47** mm), type in **48.5** mm and regenerate the assembly. **Hole 1** at the end of the shaft was created to be aligned with **DTM 1** and therefore will move together with the datum plane. This will now create the required space for the steel washers at each end.

(41) Create a new component WASHER_STEEL.PRT in Assembly mode following the instructions given in the previous Section 7.7. The dimensions are **15.5** mm internal and **24** mm external diameters, and **1.5** mm thickness.

(42) Open the new WASHER_STEEL.PRT and change its appearance. Select a standard appearance from the library, then click on **Edit Model Appearances,** in the **Appearances** pull-down menu and change it to a darker colour using the colour sliders.

(43) Use **Mirror Component** to mirror this new part and create the opposite steel washer.

(44) Check the assembly for interference, i.e. **Analysis > Global Interference > Preview**. Correct any parts with interference.

(45) Create a cross-section view to look at the connections between parts as shown in Figure 7.18.

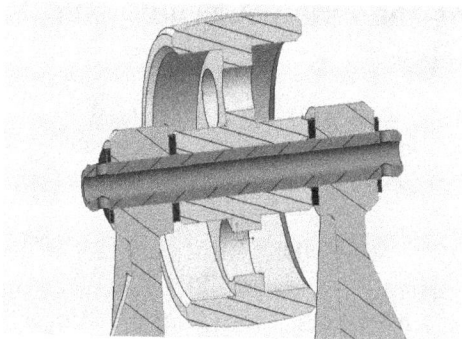

Fig. 7.18.   Cross-section of the pulley assembly.

7.9 Model Regeneration

Model regeneration has been demonstrated in the previous sections by using the **Regenerate** ( ) command. It is an operation that recalculates the model geometry feature by feature, following the feature hierarchy and Parent–Child relationship. It takes into account all changes such as dimensional changes, introduction of new features, new parts created in Assembly mode or other modifications that have been made since the last saving or opening of the model. The **Regenerate** command can reveal bad geometry or bad features, broken Parent–Child relationships, missing parts from the assembly, missing references, circular references and other problems. Features that fail to regenerate or missing components in the assembly will appear in red colour in the corresponding Model Tree. Indication of bad geometry will be a red or yellow flag that will appear at the bottom of the screen in the message area. Also, a window called **Notification Centre** will open with a list of the problems.

It is good practice to **Regenerate** the part or assembly model every time a change has been made. This way, the user can identify how specific changes might affect the model integrity. It is important to detect any problems early and resolve the failures at the time of their appearance. It is very difficult to repair a model with a chain of failures, for example several loops of circular references when working in Assembly mode.

7.10 Creating Another Part in the Pulley Assembly

The spindle is fixed to the pulley bracket by means of two split pins (cotter pins) that are inserted in the holes at the ends of the spindle. The Split pin part will be created in Part mode using the **Sweep** command along a sketched trajectory.

(46) Start a new part, i.e. **File > New** and name it SPLIT_PIN.
(47) Click on the **Sketch** and draw the sweep trajectory using the dimensions shown in Figure 7.19 (left) to create **Sketch 1** feature.
(48) Select the **Sketch 1** feature (as trajectory) and then click on the **Sweep** icon to activate. From the **Sweep** dashboard, click on the **Create or edit sweep section** icon ( ) and draw the section using the dimensions shown in Figure 7.19 (right).

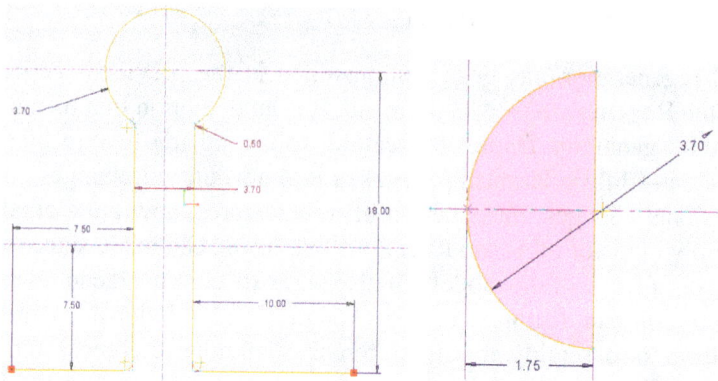

Fig. 7.19.   Trajectory (left) and section (right) dimensions in millimetre for the Sweep command (SPLIT_PIN.PRT).

Fig. 7.20.   The SPLIT_PIN.PRT part.

(49) Click on the green tick (✔) to close the **Sweep** feature.

(50) Save the part. Figure 7.20 shows the Split pin part.

(51) Switch the Active window back to PULLEY_ASSEMBLY.

(52) Click the **Assemble** icon and select the SPLIT_PIN part.

(53) In the Graphics area, select **DTM 1** datum plane (through the axis of the hole at the end of SPINDLE.PRT) as assembly reference, then select the corresponding datum plane in SPLIT.PIN (component reference) and choose the **Coincident** constraint. In addition, select two pairs of horizontal and vertical datum planes (thought the pulley

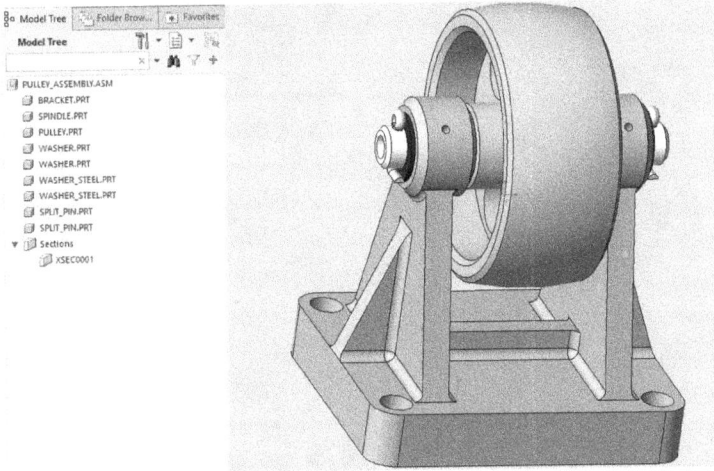

Fig. 7.21.   The final of PULLEY_ASSEMBLY.

central axis) as references for two more **Coincident** constraints until the part is 'Fully constrained'.

(54) Use **Mirror Component** command following the instructions from Section 7.7 to add another Split pin part at the opposite spindle end.

(55) Check for interference, **Analysis > Global Interference > Preview**, and correct if required. Save the assembly, i.e. **File > Save**.

Congratulations! You have completed your first assembly model PULLEY_ASSEMBLY.ASM, as shown in Figure 7.21.

Chapter 8

Mould Design

8.1 Introduction

The process of mould design (or mold according to the American spelling) is realized using a module within Creo **Manufacturing** called **Mold cavity.** This module contains special tools that to create a mould assembly model consisting of a reference part (model), mould core, mould cavity and other components such as ejector pins and standard components from a mould catalogue.

Several manufacturing technologies involve a mould as a main tool to shape a liquid (molten metal, ceramic slurry) or pliable raw material (molten plastic, glass, ceramic) in order to produce large volumes of parts. Some of the most popular of these technologies are injection moulding, blow moulding, thermoforming, sand casting, lost wax casting, powder metallurgy, spin casting, ceramic manufacturing, etc.

The **Mold cavity** provides all tools needed to design a mould model linked to the <u>reference part</u> that is supposed to be moulded.

However, this lesson will be focused on design of moulds for plastic injection moulding. The reader will learn how to import a part designed for moulding, create a set of moulds (core and cavity), modify and analyse the mould model.

Aims:
• To introduce the **Mold cavity** module and learn the mould design process;

189

- To understand the basic steps in creating, modifying and analysing a set of moulds.

Outcomes:
At the end of this lesson, the reader should be able to:

- Import a part (reference part) in the **Mold cavity** module;
- Apply shrinkage to the reference part;
- Create the mould block (workpiece) and parting surfaces;
- Split the workpiece into mould volumes, extract the mould volumes and convert them into solid models;
- Examine the mould models for correct draft angles;
- Create a test shot and simulate the mould opening;
- Create cooling channels and a gating system (runners, sprue and gate);
- Add ejector pins.

8.2 Mold Cavity Tutorial

Mold cavity module has tools to create a mould model as well as to modify and analyse the mould tool components' fitness for mould manufacture.

8.2.1 *Mold cavity workflow*

The **Mold cavity** workflow consists of the following steps:

- Preparation of the reference part (model);
- Import of the reference part. This step creates a copy of the part, but with a new name XXX_REF.PRT, containing the part geometry as a single feature;
- Create the workpiece (mould block volume) that will enclose the reference part and produce the mould and cavity;
- Create the parting lines and Parting Surfaces that are used to split the workpiece;
- Split the workpiece into two or more mould volumes;
- Extract the mould volumes and create mould components (solid versions of the workpiece volumes);

- Inspect the mould components and analyse draft angles to ensure that the moulded parts can be ejected;
- Simulate a test shot and mould parts opening;
- Create waterlines, gating system and ejector pins.

The mould model created in the **Mold cavity** module is in fact an assembly (*.ASM) containing all mould tool parts (components). The first parts to be assembled are the reference part and workpiece. After that, all other mould components are developed and then saved in the Working Directory. Despite being an assembly model, whenever the mould assembly is opened, it will automatically run the **Mold cavity** module.

8.2.2 *Reference part* (*model*)

Before starting this tutorial, open the **Windows Explorer** or **My Computer** and perform the following:

- Create a subdirectory C:\USER\MOLD_CAVITY on your PC.

    (1) Run Creo and create a new part LID.PRT (in <u>Part mode</u>) using a Metric (mm) template. All dimensions of this part are shown in Figure 8.1. Create the model utilizing revolve, extrude, hole, draft, round and other features from the previous chapters. Add **1** degree draft angle to all vertical walls to facilitate the mould opening. Generally, all surfaces that might have undercuts with respect to the pull direction should be modified to ensure easy ejection of the moulding.

- Save the Lid part into C:\USER\MOLD_CAVITY directory.

8.2.3 *Reference part preparation*

    (2) Start Creo (unless it is running).
    (3) Set up the Working Directory, i.e. **File > Manage Session > Select Working Directory,** and then select C:\USER\MOLD_CAVITY folder. This will ensure that all new models will be saved in one place.
    (4) To avoid potential problems in the mould design, check the LID. PRT model integrity by opening it in <u>Part mode</u> and then perform

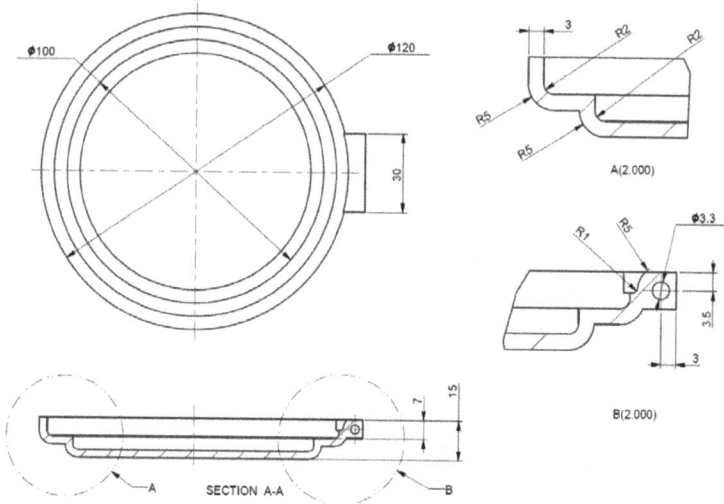

Fig. 8.1. Drawing of the reference part LID.PRT.

File > Prepare > Model Check Interactive. Investigate and resolve if there are 'buried features' in the report (see Chapter 6, Section 6.17).

(5) Check that the **enable_absolute_accuracy** configuration parameter is set to **yes.** To do this, select **File > Options > Configuration Editor.** Click on **Find** icon (at the bottom of the window) and type *enable_abs* in the **Type keyword** slot. Click on **Find now**, select **enable_absolute_accuracy** and in the **Set Value** slot change the value from **no** to **yes.** Click on **Add/ Change**, then **Close**, and **OK.** Select **Yes** and **OK** to save to a new configuration file.

(6) Select **File > Prepare > Model Properties** (Figure 8.2).

(7) Find the **Accuracy row** and click on **change.** In the window **Accuracy**, change **Relative** to **Absolute**, and then enter a very small value, e.g. **0.0001**, and click on **Regenerate Model.** Change to the smallest value that allows successful model regeneration.

(8) Click on **Close** to finish editing the model properties.

It is a good idea to **Hide** some datum features such as planes, axes, surfaces, curves, scan curves, etc. for a clearer model display (see Chapter 3). Save the visibility status using the **View** tab > **Visibility** group > **Save**

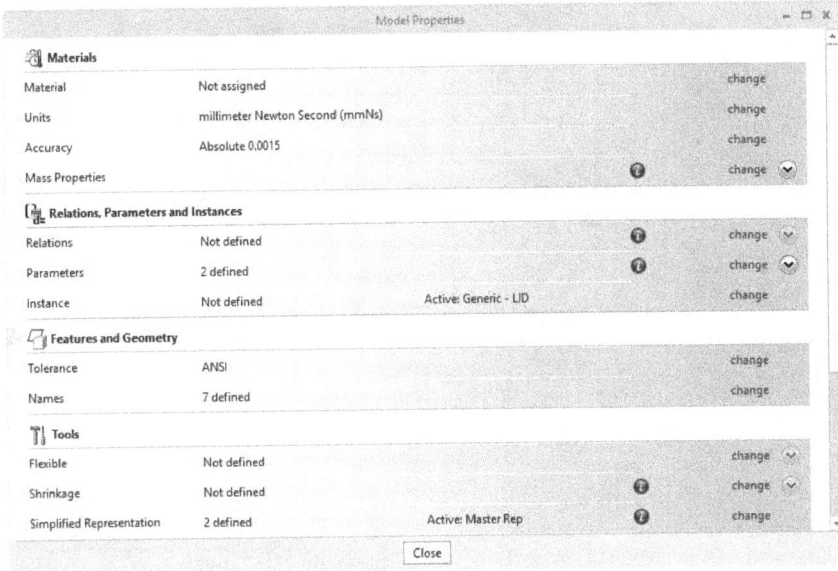

Fig. 8.2. Model Properties.

Status icon ( ). Finally, save the part (LID.PRT) in the Working Directory e.g. **File > Save** ( ).

8.3 Start Mould Design and Import the Reference Model

(9) Create a new file by selecting **New** ( ), or **File > New**. In the **New** dialogue window, click on the **Manufacturing** and **Mold cavity** small circles (buttons) to activate as shown in the Figure 8.3. Type a model name, i.e. **mold**, remove the tick from the **Use default template** check box and click on **OK**.

(10) In the **New File Options** window, select **Empty** and then **OK** to accept. In this case, the reference model will bring its template with the corresponding datum planes and units (mm). Alternatively, instead of empty, you can select the **mmns_mfg_mold** template.

(11) The **Mold** ribbon interface opens. The Model Tree displays the mould model name MOLD.ASM (Figure 8.4).

(12) From the **Model Properties,** check the **Accuracy** and make sure that it is set to **Absolute**. If not, then perform the steps from the

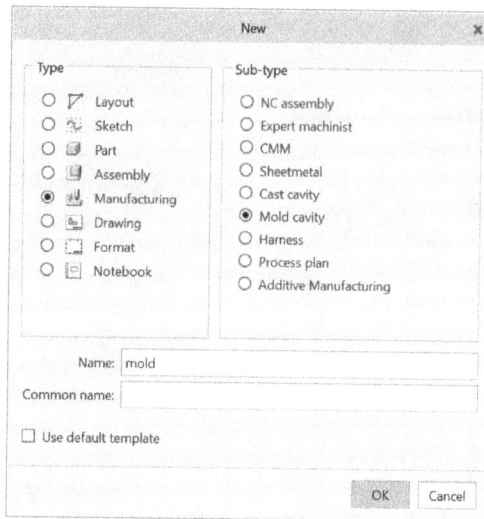

Fig. 8.3.   New file window.

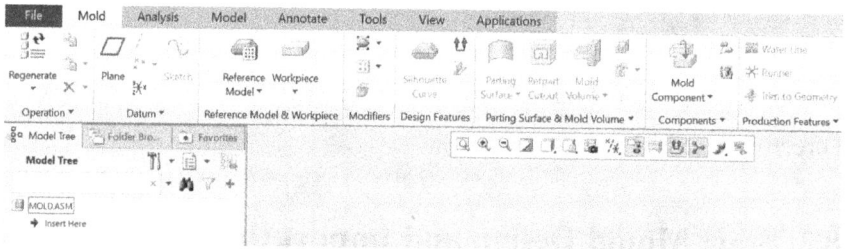

Fig. 8.4.   The empty MOLD.ASM file with an empty Model Tree.

previous section, change to **Absolute** and enter the smallest possible value.

(13) Continue creating the mould model. From the **Mold** tab, **Reference Model & Workpiece group**, click (LMB) on the **Reference Model** icon ( ) and choose **Assemble Reference Model**. Select the LID part and click on **Open** as shown in Figure 8.5.

Important: The default System of Units is Imperial (inch, pound) unless it was set to Metric during the software installation (see Chapter 1).

(14) To check the units, go to **File > Prepare > Model Properties**. If the **Units** row shows **inch lbm Second**, then click on **change** and

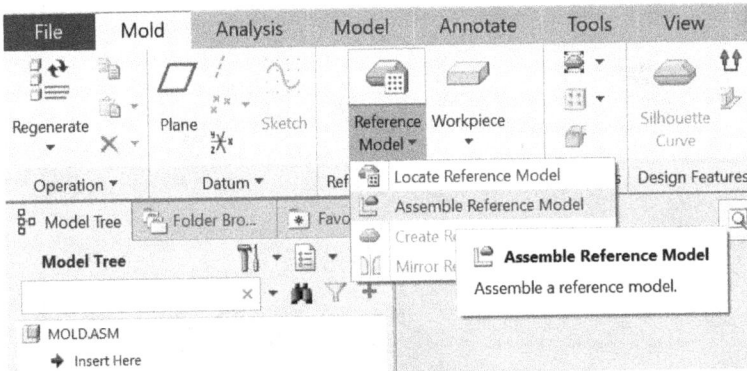

Fig. 8.5.   Locate the Reference Model.

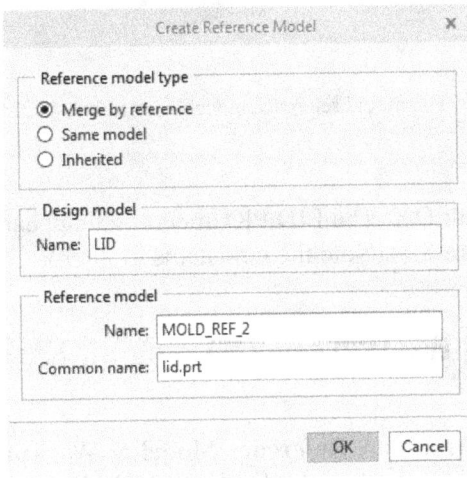

Fig. 8.6.   Create Reference Model — Merge by reference.

from the **Units Manager** window select **millimetre Newton Second**, tick **Include submodel** and then click on **Set**. In the new window **Changing Model Units,** keep the pre-selected values. Select **Convert dimensions**, click on the **OK** and then **Close** the **Units Manager**.

(15) In the **Create Reference Model** dialogue window, keep **Merge by reference** selected and click on the **OK** icon as shown in Figure 8.6.

(16) A **Warning** window appears prompting to confirm the absolute accuracy which the assembly inherits from the reference model.

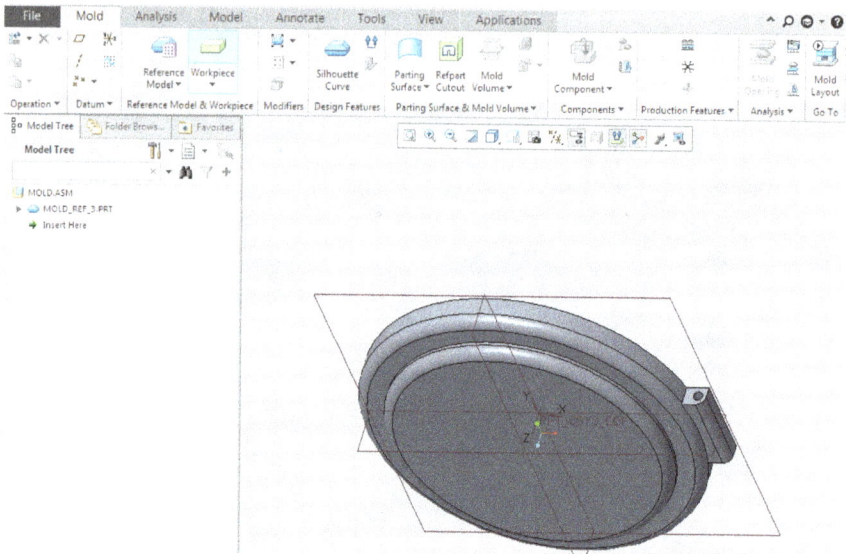

Fig. 8.7.   The mould reference model inserted into MOLD.ASM.

Accept with **OK**. The LID.PRT model will appear in the Graphics area (Figure 8.7). Note the new items in the Model Tree.

8.4 Creating the Workpiece Using an Extrude Feature

(17) From the **Mold** tab, **Reference Model & Workpiece** group, click on the down arrow at the bottom of **Workpiece** icon and select **Create Workpiece** (**Mold > Create Workpiece**). In the **Create Component** window, keep the defaults **Part** and **Solid** and type BLOCK in the **Name** slot. Click on **OK** to accept. In the **Creation Options** window, select **Create feature** and click **OK**.

(18) From the **Mold** tab, **Shapes** group, select the **Extrude** icon ( ). The Sketcher dashboard appears at the top of the screen.

(19) Click on the **Placement** tab, and then select **Define** to open the **Sketch** window for the sketch and reference planes selection. Pick the datum plane that goes through the model axis of rotation as

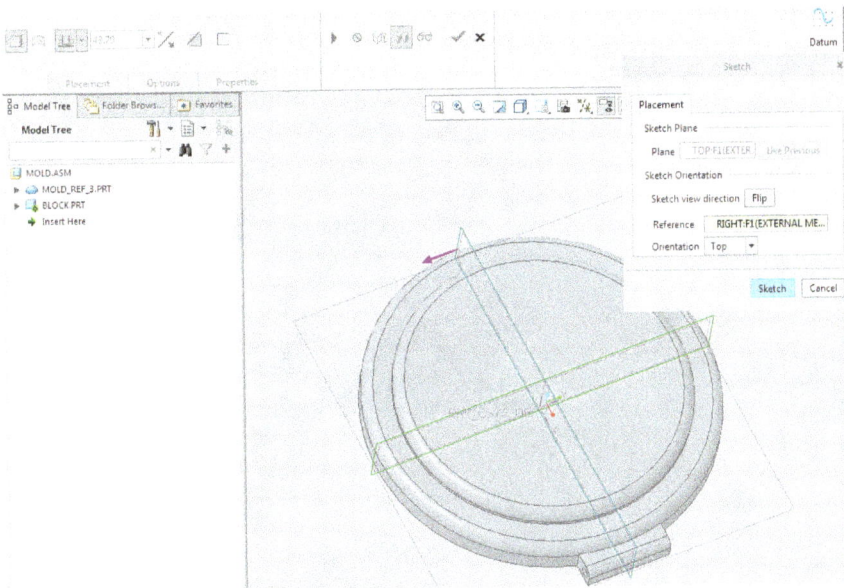

Fig. 8.8.   Sketch Plane selection.

Sketch Plane (Figure 8.8). When selected, it is marked with a magenta arrow normal to the plane. Next, select a plane normal to the sketching plane as **Reference** (plane).

(20) Note that the **Sketch** dialogue window slots contain the names of the selected datum planes. Click on **Sketch** icon to accept and proceed.

(21) The **References** window appears. Select at least two references (datum planes or edges normal to each other) or the coordinate system in the middle. Click on **Solve** and if the result is **Fully Placed**, then click on **Close**.

ⓘ Workpiece dimensions: Mould blocks (inserts) are subjected to extremely high injection pressures during the IM process. Therefore, the mould inserts need to be rigid enough to prevent deformation and squirting of molten material from the joint faces under pressure. In addition, the mould inserts should fit into standard base blocks (bolsters) that are attached to the moulding machine. For simplicity, in this example, the workpiece size is defined to be three times larger than the reference part, i.e. **380 × 380 × 50** mm.

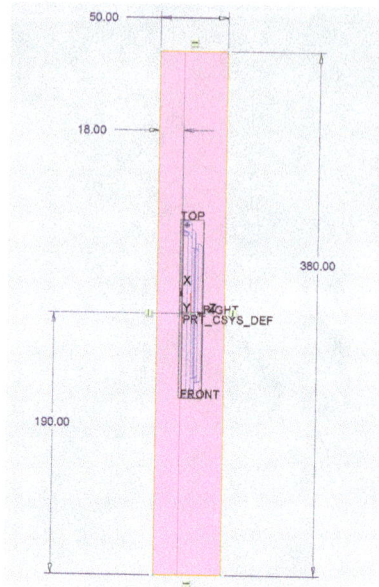

Fig. 8.9.   Feature cross-section.

(22) Click on the **Sketch View** icon (⟳) to align the view to the screen
unless you have set up the parameter **sketcher_starts_in_2d** to
yes. If you have not done it yet, then do it now (see Chapter 6,
Section 6.2). Select **Corner rectangle** (▢) tool and draw a rectangular block symmetrically around the reference model.

(23) Dimension the sketch as shown in Figure 8.9, i.e. rectangle of **380
× 50** mm. Draw a centreline and create symmetrical constraints to
reduce the number of dimensions.

(24) Press the green tick (✔) to close the Sketcher.

(25) Re-orientate the view (MMB+drag) to be able to see the extrude
depth. Select the **Extrude both sides** option (⊟) from the **Extrude**
dashboard and enter the overall depth of **380** mm.

(26) Click on the green tick (✔) to save and exit. The workpiece
(BLOCK) is shown in Figure 8.10.

✎ The BLOCK.PRT model can be modified or redefined as follows:
expand the Model Tree to view the features, click on the **Extrude** feature
and select **Edit Dimensions** or **Edit Definition**.

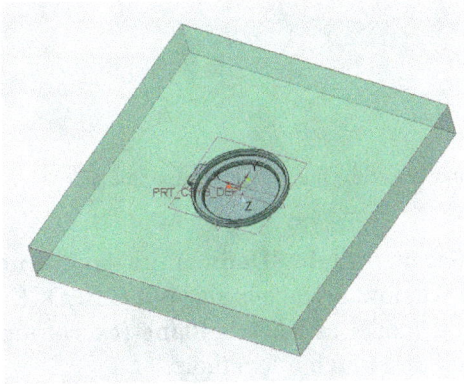

Fig. 8.10.   Workpiece (BLOCK.PRT).

During the BLOCK.PRT creation procedure, a green dot, next to its name in the Model Tree (▶ ⬜ BLOCK.PRT), indicates that this part is active or under editing. Once the procedure is closed, the green dot should disappear. If the BLOCK.PRT is still active, then the user should finish and close the **Extrude** or activate back the MOLD.ASM assembly by selecting the MOLD.ASM name in the Model Tree, and then click on the **Activate** (◆) icon from the mini menu.

(27) **Save** (💾 Save) the model.

8.5 Creating the Parting Surface

Parting surface(s) divide the BLOCK model into <u>mould core</u> and <u>mould cavity</u>. The selection of correct parting surface is essential for the mould design. It ensures successful mould opening and moulding extraction (ejection). In this relatively simple and straightforward example, a single flat parting surface will be created.

(28) To create the parting surface, click on the **Parting Surface** (⬭) icon, (**Mold > Parting Surface**), from the main ribbon. Next, select the **Extrude** (Extrude) command to create an extruded surface.

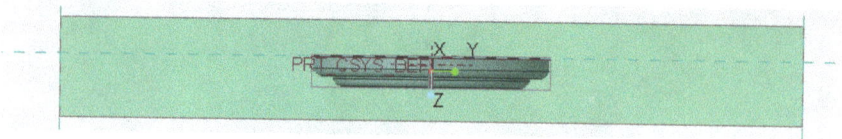

Fig. 8.11.   Three reference lines to aid the Extrude of a flat parting surface.

(29) Click on **Placement tab > Define** from the **Extrude** dashboard. In the **Sketch** window, select any side of the BLOCK that is not parallel to the lid model as **Sketch Plane** (except top or bottom surfaces). Click on **Sketch** to continue.

(30) In the **References** window, select the horizontal top edge of the lid and both vertical edges of the BLOCK (Figure 8.11). Click on **Solve**, and then **Close** if the result is **Fully Placed**.

(31) Select the **Line Chain** tool (⌢) and sketch a line. Draw a horizontal line snapping the first point to the intersection of the left vertical reference with the horizontal reference and the second point to the right vertical reference. Click the MMB to stop and close the Sketcher. In order to view the parting surface direction, rotate the model view. It is ok if the parting surface is outside the BLOCK.

(32) From the dashboard, select **Extrude to selected** option ( ) and then click on the BLOCK side parallel to the sketching plane to extrude in one direction. Go to dashboard **Options**, select **Side 2** and then click on the opposite BLOCK side (the second direction). The parting surface should split the entire block, as shown Figure 8.12.

(33) Click on the green tick (✔) to save and close.

(34) Complete the parting surface and select the green tick (✔) from the **Controls** group (main ribbon) to close (Figure 8.13).

8.5.1 *Redefining the parting surface*

(35) The parting surface is a feature that belongs to the mould assembly and can be redefined using **Edit Definition** command. By default, it may not be displayed as a feature in the Model Tree. In order to reveal it, click on **Settings** ( ) at the top right of the Model Tree, and then select **Tree Filters**. Tick the **Features** box (**Model Tree Items** window). Also, to display any suppressed features or parts,

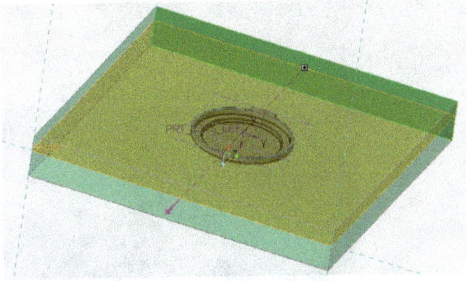

Fig. 8.12.   The extruded flat parting surface.

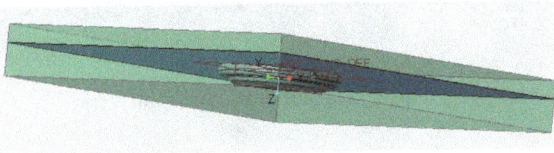

Fig. 8.13.   Completed parting surface.

tick the **Suppressed objects** box in the same window. (Refer to Chapter 3, Section 3.5 for more details.) Click **OK** to close.

(36) Now all features, including the <u>parting surface</u> **Extrude 1**, should be visible in the Model Tree. If you need to redefine it, then click on it and select **Edit Definition** from the mini menu.

(37) Click on **File** > **Save** ( ) to save the model after modifications.

8.6 Splitting the Workpiece

This section describes how to split the workpiece (BLOCK) using the parting surface and mould reference part.

(38) Click on the **Refpart Cutout** icon ( ) in the ribbon to open the dashboard, (**Mold** > **Refpart Cutout**). Click on **References** and check if BLOCK and MOLD_REF parts are pre-selected. Click the green tick ( ) to accept.

(39) To split the workpiece into volumes, click on the **Mold Volume** ( ) arrow to open the drop-down list and select **Volume Split** ( ), (**Mold** > **Volume Split**).

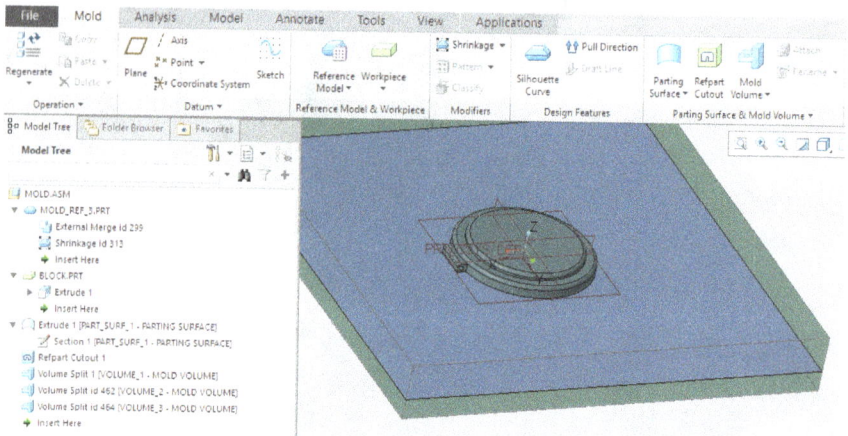

Fig. 8.14.   Initial volume split.

(40) In the **Volume Split** dashboard, click on **References** to open and under **Split Surfaces** select the <u>parting surface</u> from the Graphics area or from the Model Tree. Click on the green tick (✓) to accept.

Note the volume split list in the Model Tree (Figure 8.14). The block splits into several parts, and you might have some unattached volumes called 'islands'. The aim is to design a two-part mould Core-Cavity, also known as Male–Female configuration.

(41) To attach the 'islands' to the relevant mould part, click on the **Attach** icon ( ). The **Search Tool** dialogue window opens. A list of all available volumes (Quilts) is shown in the bottom left corner, also in the Model Tree. Click on each volume (quilt) in the list to highlight, inspect it on the screen, and decide which volumes to attach together.

(42) Select the top quilt, VOLUME_1. It is the cavity (female) part, which looks complete. No attachment is needed.

(43) Select the second quilt, VOLUME_2. It is supposed to be the core (male) part, but it looks incomplete without the internal lid area.

(44) Select the third quilt, VOLUME_3. It is the missing internal area of the core. This volume should be attached to VOLUME_2. To do that, select VOLUME_2 and click on the double arrow icon >> to copy it to the right window and then **Close**.

(45) Repeat this with VOLUME_3, and when you click on **Close** it will disappear from the list. Thus, only VOLUME_1 and VOLUME_2 (joined with VOLUME_3) remain. Continue the above procedure if there are more unattached items until only two (core and cavity) remain in the list of available volumes. Click on **Close** to finish.

Notice the new features in the Model Tree.

8.7 Extracting the Mould Volumes from the Workpiece

At this stage, the mould volumes are features of merged surfaces and volumes. They need to be converted into solid parts.

(46) To convert the volumes into solid parts, go to **Mold > Mold Component** ( ) in the main ribbon, click on the down arrow (not the icon itself) and select **Cavity Insert** ( ). In the **Create Mold Component** window, select both **VOLUME_1** and **VOLUME_2**. Press (CTRL + hold) to select both volumes. Click **OK** to finish. Two new parts should appear now in the Model Tree (Figure 8.15).

(47) The user could open separately and examine these two parts to make sure that they are correct. Select VOLUME_1 in the Model Tree and click on **Open** from the mini menu. Repeat this with VOLUME_2.

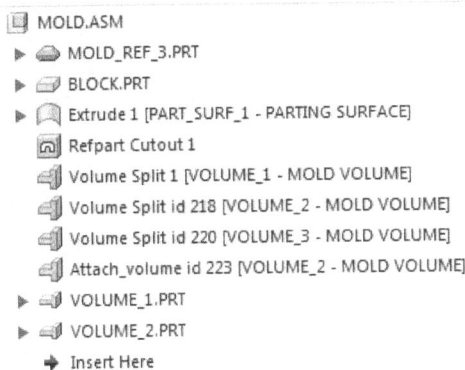

```
   MOLD.ASM
 ▶  MOLD_REF_3.PRT
 ▶  BLOCK.PRT
 ▶  Extrude 1 [PART_SURF_1 - PARTING SURFACE]
    Refpart Cutout 1
    Volume Split 1 [VOLUME_1 - MOLD VOLUME]
    Volume Split id 218 [VOLUME_2 - MOLD VOLUME]
    Volume Split id 220 [VOLUME_3 - MOLD VOLUME]
    Attach_volume id 223 [VOLUME_2 - MOLD VOLUME]
 ▶  VOLUME_1.PRT
 ▶  VOLUME_2.PRT
 ➜  Insert Here
```

Fig. 8.15.   Solid volume parts in the Model Tree.

Fig. 8.16.   Cavity VOLUME_1 (left) and core VOLUME_2 (right).

(48) These two parts should look like Core-Cavity components (Figure 8.16). If they do not, and look like two flat plates, then most likely this is a result of a wrong island combination. In that case, the user should redo the island attachment procedure.

8.8 Opening the Mould Model

Opening the mould model would allow a quick check of the design configuration and interaction of mould components. Proceed as described in the next steps.

(49) Hide some parts for clearer view. Click on BLOCK.PRT in the Model Tree and select **Hide** (✖). Also, hide the parting surface and then save the visibility status (Section 8.2.3).

(50) To create mould opening, click on the **Mold Opening** icon (🖫) from the main ribbon (**Mold > Mold Opening**). This will open **MOLD OPEN** window menu. Select **Define Step > Define Move**. From the Model Tree select a mould part, for example VOLUME_1, and click **OK** to confirm. Following the instructions in the message area (bottom of the screen), select a direction to move — any datum plane or an edge. For example, select the top flat surface of VOLUME_1 (see Figure 8.17). Enter **100** mm movement and click the green tick icon (✔) to confirm. Click on **Done** and VOLUME_1 will move out.

(51) Define a step for VOLUME_2 to move in the opposite direction. Use minus sign (**-100**) to specify the opposite direction. The opened mould is shown in Figure 8.18.

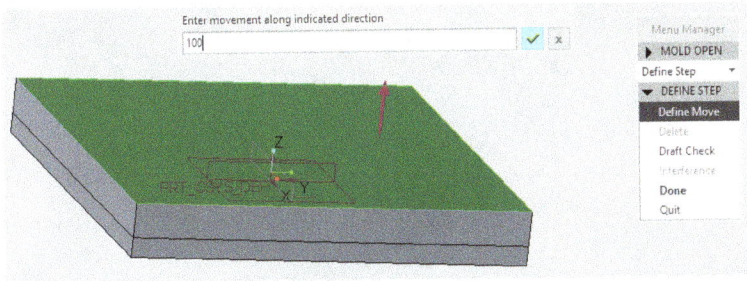

Fig. 8.17.   Volume opening direction.

Fig. 8.18.   Mould opening.

If a mistake is made, delete the steps. Click on **Delete All** from **MOLD OPEN** or **Delete** from **Define Move** windows and start again.

(52) To see an animated exploded view, click on **Explode** and under **STEP BY STEP** select **Open Next** (step by step) or **Animate All**. To close, click on **Done/Return**. Click (LMB) again to close the mould.

(53) Select **File > Save** (🖫 Save) to save the model.

8.9 Draft Angle Analysis

In order to prevent problems with the mould opening and moulding ejection, it is vital to make sure that correct draft angles have been applied to the part surfaces.

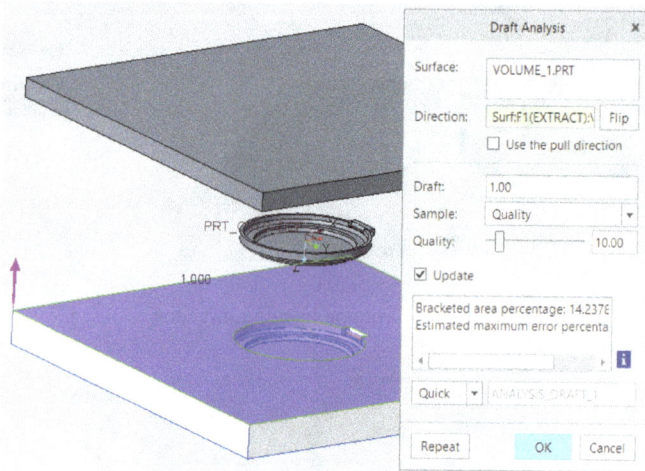

Fig. 8.19.   Draft Analysis dialogue window (VOLUME_1).

(54) To perform this analysis, first open the mould (unless it is already opened). Click on the **Mold Opening** icon ( ) to open and change the **Display Style** (Graphics toolbar) to **Wireframe**.

(55) Select **Analysis** tab to switch to the analysis ribbon interface. Select the **Draft** icon ( ), (**Analysis > Draft**). In the **Draft Analysis** dialogue window (Figure 8.19), remove the thick mark in front of **Use the pull direction** (unless the pull direction indicated on the screen is correct) and change the **Draft** angle to the smallest angle that is allowed, e.g. **1** degree.

(56) Click inside the **Surface** slot to activate and then select VOLUME_1 PRT name in the Model Tree. Click inside the **Direction** slot to activate and select the surface with a normal vector pointing toward the pull direction. This is the direction of pulling the moulding out of the mould part. If necessary, click on **Flip** to change the direction. Figure 8.20 shows the result of the draft analysis.

The colours in in Figure 8.20 represent ranges of graft angles shown on the left side. Almost all surfaces are displayed in blue, which means that they have drafts larger than **1** degree, which is correct. Only one surface is grey/white, which is practically a **0** degree draft. This could hinder

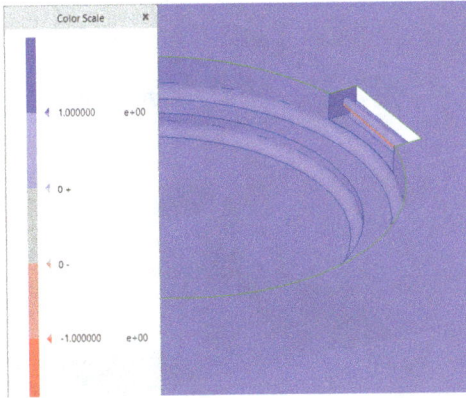

Fig. 8.20.   Draft analysis results for VOLUME_1.

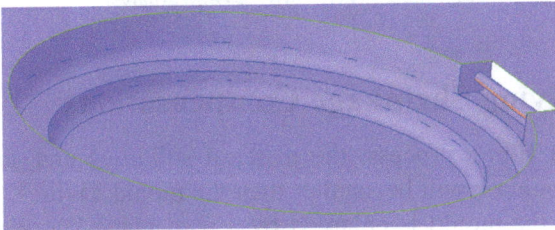

Fig. 8.21.   Missing drafts (white colour) and undercut (red colour).

the ejection. On a closer look in Figure 8.21, the hinge semi-cylindrical surface is pink-red. This is a negative draft also called 'undercut', which would make the ejection impossible.

(57) Click on the **Repeat** icon (**Draft Analysis** window), change the setup and analyse VOLUME_2 part (Figure 8.22).
(58) Alternatively, each mould insert can be opened in <u>Part mode</u> and then subjected to draft analysis, i.e. **Analysis > Draft** ( ).

Important: Any problems with the draft angles identified in the analysis <u>must be fixed</u> in the part model. To do that, **Open** the LID.PRT, add necessary drafts and remove the undercuts. **Save** the LID.PRT. Return back to the MOLD.ASM and **Regenerate** the whole assembly. Repeat the draft analysis until all draft problems have been fixed.

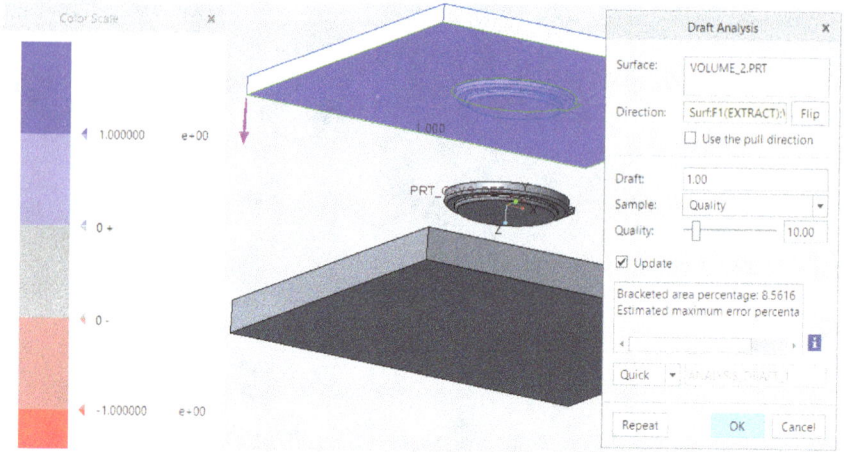

Fig. 8.22.   VOLUME_2 Draft Analysis.

8.10 Applying Shrinkage Compensation

After the IM filling process, the material solidifies and shrinks. As a result, the moulding will be smaller than the mould cavities. To compensate for this, the mould is enlarged by a percentage know as shrinkage factor or patternmaker's shrinkage.

(59) To apply the shrinkage factor, select **Mold** tab > **Modifiers** > **Shrink by scaling** (⬚) icon. Next, select the mould co-ordinate system and enter the shrinkage coefficient/ratio as isotropic or non-isotropic (X, Y and Z direction separately), whichever is appropriate. In this example, an average shrinkage factor of **1%** (type **0.01** value) isotropic is used. For better accuracy, find the exact value relevant to the plastic material.

(60) Click on the goggles icon (⬚) to preview, and if the result is correct click the green tick (✔) to accept. You may need to regenerate the assembly model after that.

⬚ Note, unlike the **Scale Model** command (from **Model** tab > **Operations**), which cannot be reversed, the **Shrinkage** is a feature that can be modified with **Edit Dimensions** or **Edit Definition**. It is located below the reference part in the Model Tree.

It is advisable to measure some dimensions from the reference part, mould core-cavity, and to check that they are slightly larger than the corresponding dimensions from the LID.PRT part after the shrinkage application.

8.11 Test Shot

For further visual analysis, the user can create a test shot part (moulding). It represents the liquid (molten) state of the moulding before solidification, and it should be slightly larger than the LID.PRT part. You can measure some dimensions from both to make sure that the test shot part is larger, adopting the shrinkage applied to the LID.PRT.

(61) From the **Mold** tab, **Components** group, click on the **Create Molding** icon ( ), type a name, e.g. TEST, and then click on the green tick ( ) twice. The test shot will appear in the Model Tree. Open the TEST.PRT part to examine it.

8.12 Modifications to Mould Volumes: Creating a Slider (Mandrel)

The draft angle analysis identified an undercut in the area of the hinge, pink-red area in Figure 8.21. A possible way to overcome this problem is to create another component of the mould tool called a slider. It is also called a 'mandrel' in the case of a slider with cylindrical cross-section. This part slides inside the mould tool, perpendicular to the mould opening direction, before the injection of molten plastic. After solidification and before the moulding ejection, the mandrel slides out, thus releasing the undercut area.

(62) In order to view the mould's internal structure, switch the display style to wireframe ( ). Click on the **Mold > Mold Volume** (Mold Volume▾) down arrow to open the drop-down menu and select the **Mold Volume** ( ) icon, in **Parting Surface & Mold Volume** group. **Edit Mold Volume** ribbon opens with tools for mould volume modification.

8.12.1 *Designing a slider*

There are <u>two possible methods</u> to design a slider: automated and manual.

Automated method: Creo will automatically detect all trapped volumes that may need sliders.

(63) In **Edit Mold Volume** ribbon, select the **Slider** (Slider) command, **Volume Tools** group. In **Slider Volume** dialogue window, define the **Pull Direction** (mould opening) by selecting the top (or bottom) mould side. The direction is shown as a red arrow in Figure 8.23.

(64) Click on **Calculate Undercut Boundaries**. The system automatically identifies the undercuts (quilts) and lists them in the **Exclude** area. Select only those relevant to the slider(s) that are to be created. Preview these quilts by hovering with the mouse cursor over the items in the list as they flash in dark red colour. For better visibility, select a quilt name and click on the quilt icon (   ) to mesh or click on shade (   ) to shade the corresponding quilt (Figure 8.24).

(65) In this case, both quilts are relevant as they both define the boundaries of the LID hinge hole. To transfer a quilt, select the quilt first

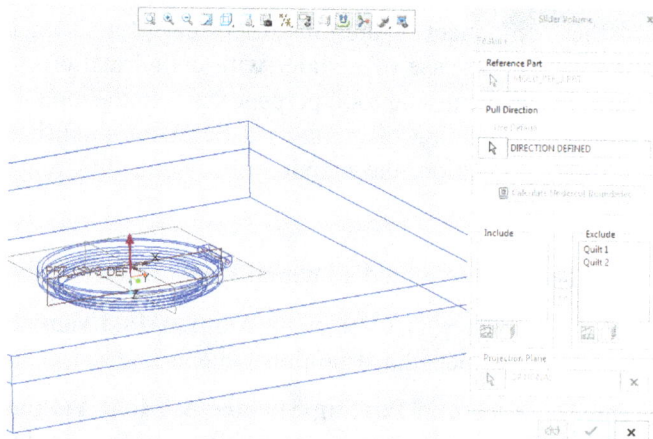

Fig. 8.23.   Designing a slider.

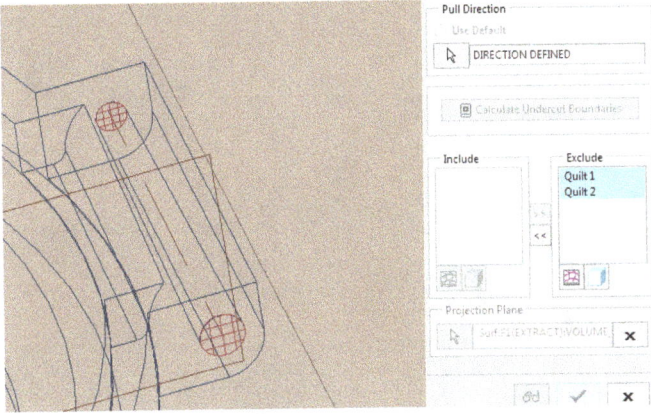

Fig. 8.24. Designing a slider — the hole in the LID with undercut.

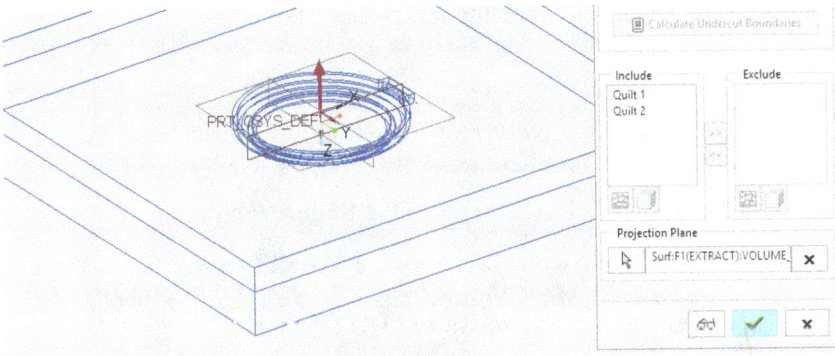

Fig. 8.25. Designing a slider.

and then click on the left (or right) double arrow (⧉) icon between **Include** and **Exclude** areas.

(66) The slider body includes the trapped volume (shown in Figure 8.24) and the volume of the extension to the mould boundary. To create the slider body, click on the arrow in the **Projection Plane** (at the bottom of the current window) and select the side surface of the mould from where the slider will slide in or out (green colour in Figure 8.25).

If the hole in LID.PRT has a draft, then select the good side that allows the slider to be pulled out. In this case, Quilt1 and Quilt2 are the

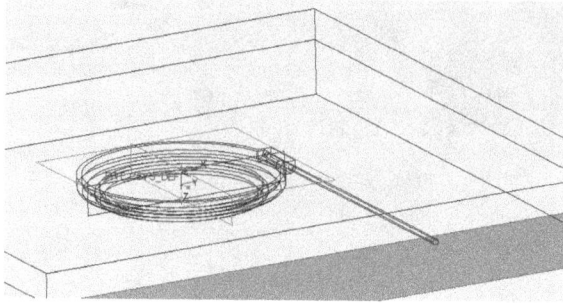

Fig. 8.26.   Designing a slider.

larger and smaller circles creating a conical (drafted) surface in between. Choose the extension direction towards the larger circle, so that the body (shank) of the slider will have the larger diameter.

(67)   Click on the goggles icon (👓) to preview the result. If the outcome is correct (Figure 8.26), click the green tick icon (✔) to close the **Calculate Undercut Boundaries** window, and then click on the next icon (✔) to close the **Slider Volume**.

(68)   Select the **Shade** icon (shade) from the **Visibility** group (the left-hand side of the **Edit Mold Volume** ribbon) to change the display style, and view the slider geometry.

(69)   If the result is correct, click the green tick (✔) to save and close the **Edit Mold Volume** ribbon.

(70)   Extract the slider volume in the same way as previously described in Section 8.7. It will appear at the bottom of your Model Tree as SLIDER.PRT (shown in Figure 8.27 as an opened part).

Note that only the slider tip, which is in contact with the molten material, is tapered or drafted. The rest of the slider is cylindrical for easy fit with the mould insert.

Manual method: Sometimes due to geometric complexity, the automatic detection of slider quilts may not work. The alternative is to directly use solid modelling tools and create a slider.

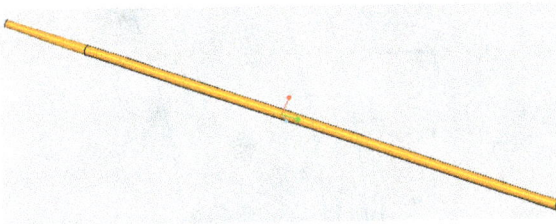

Fig. 8.27. Extracting the slider volume.

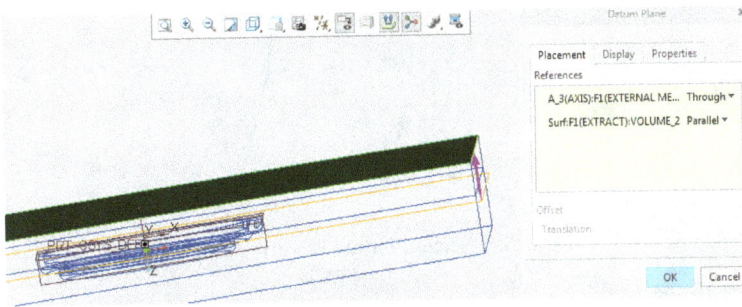

Fig. 8.28. Datum plane for sketching (created 'on the fly').

(71) Switch back to the **Edit Mold Volume** ribbon and instead of **Slider,** select the **Revolve** ( ) command. Click on the **Sketch** icon ( ), in **Datum** group on the right-hand side of the **Revolve** dashboard, and then select the **Plane** icon ( ) to create a plane for sketching. (This plane should be through the axis of the hole with the under-cut.) In order to create a plane through an axis, select the axis of the hole first, press (CTRL + hold), and then select the top surface of the mould. The selected references appear in the **References** area, **Datum Plane** window. Change the suggested **Offset** option to **Parallel** (to the selected reference surface) as shown in Figure 8.28. Click on **OK** to finish and close the **Datum Plane** window.

(72) In the dashboard, click on the **Resume** icon ( ) and, in the next window, **References,** select two perpendicular planes or the coordinate system (a faster way). Click on **Solve** and if the status is **Fully Placed** then click on **Close**. Use the **Centreline** ( ) tool, **Datum** group, and draw a datum centreline for the **Revolve**

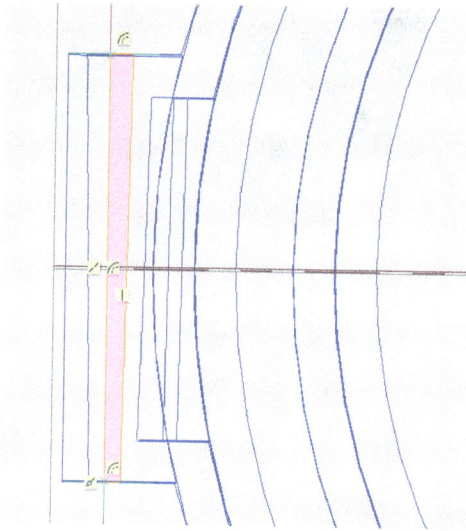

Fig. 8.29.   Section for the Revolve feature.

feature. Make sure it is coincident with the hole axis. If not, then use the **Coincident** tool (⁓), **Constrain** group, to align.

(73) Use **Project** (▯) and **Line Chain** (⌵) tools, from the **Sketching** group, to draw only half of the slider cross-section following the visible existing edges of the hole and the centreline. Use the **Corner** (⊤) tool to trim and create a closed section (see Figure 8.29).

(74) Finish the revolved section and click the green tick (✔) to close.

(75) In the **Revolve** dashboard, keep **360** degrees and if the preview is successful, click on the green tick (✔) to close the feature.

(76) Select the green tick (✔) to save and close the **Edit mold volume** ribbon. A new volume **Revolve 1** appears in the Model Tree.

(77) You have to extend the shape of the volume to the outside wall of the mould insert, similar to the automated design option. Click on the **Revolve 1** volume feature in the Model Tree and select **Edit definition**, then **Placement**, and **Edit**. Use the **Sketch** tools to redefine the section and extend the slider (see Figure 8.26).

(78) Extract the slider volume and create a solid part similar to Section 8.7.

8.12.2 *Cutting the slider volume out of the mould*

The slider volume should be cut out from all mould inserts that have interference with the slider. You can identify those parts from the **Analysis** tab > **Global Interference** check.

(79) Select and **Open** ( ) the part that intersects with the slider. In this case, it is VOLUME_1.PRT.

(80) From the **Model** tab, click on the **Get Data** down arrow and select **Merge/Inheritance**. Click on **Remove material** icon ( ) from the dashboard and then select the **Open a model...** icon ( ). In the **Open** window, select the part to be subtracted, i.e. select the SLIDER.PRT, and then click on **Open** to see it in a separate window. Also, the **Component Placement** window opens.

(81) Click on the goggles icon ( ) to view the current SLIDER. PRT position. All parts in the mould assembly should have a common coordinate system, i.e. the default. Therefore, click on the down arrow and change the **Constraint Type** from **Automatic** ( ) to **Default** ( ).

(82) If the part placement is correct (see Figure 8.30), then click the green tick ( ) and close the **Constraint Type** window and the

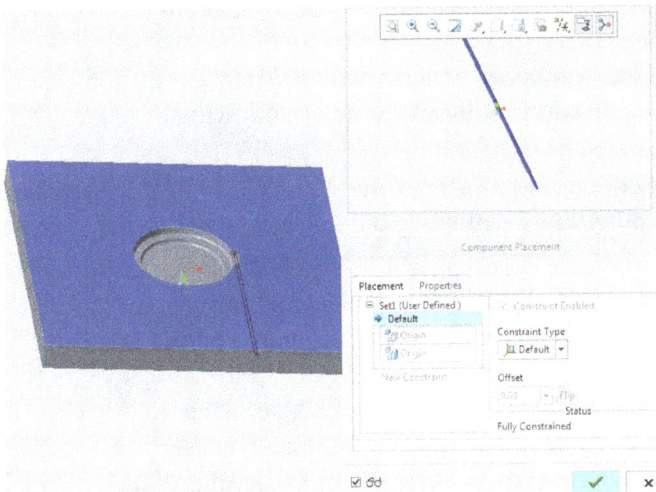

Fig. 8.30.   Cutting the slider volume out of the mould.

dashboard. Examine the modified part VOLUME_1 and observe the new feature; the volume that has been subtracted to accommodate the slider. Also, the undercut in the hinge area has disappeared.

Another way to create the above 'cut out' directly from the **Mold** tab is as follows: From the **Mold** tab > **Components** group, click on the down arrow and select **Component Operations** > **Boolean Operations**. In the new window (on the top), change **Merge** to **Cut**. In **Modified Models**, select VOLUME_1.PRT and in **Modifying Components** select the slider SLIDER.PRT. Keep the defaults and click on **OK** to finish.

(83)   Click on **File** > **Save** to save the VOLUME_1.PRT and then **Save** the MOLD.ASM.

8.13 Modifications to the Mould Volumes: Creating Cooling Channels (Water Lines)

Plastic injection moulding process is very sensitive to temperature changes, and usually a stable working temperature is required to avoid problems such as incomplete shots, cold weld lines, warping, voids, and other issues. A cooling system with channels, often called water lines, created in the main mould inserts, helps to maintain stable thermal conditions during the injection moulding. Usually, the cooling lines are drilled in a ∏-like pattern and then the inlet and outlet are tapped to fit plugs or connectors.

Creo has a number of tools within **Mold cavity** for creating water lines. The procedure is illustrated in the following example.

(84)   Click on the **Water Line** icon (Water Line), in **Mold** tab > **Production Features** group, to activate. Enter **8** mm water line diameter, and accept with the green tick (✔). **Water Line** window opens. In the **Menu Manager** sub-window, select (or **Setup New**) sketching plane. (Remember to read and follow the messages at the bottom of the screen.)

(85)   The channels are created in a single plane. In this case, create an offset datum plane from an existing plane surface. First, switch the display style to **Wireframe**. Click on the **Plane** (▱) icon from **Mold** > **Datum** group. The new window **Datum Plane** prompts for

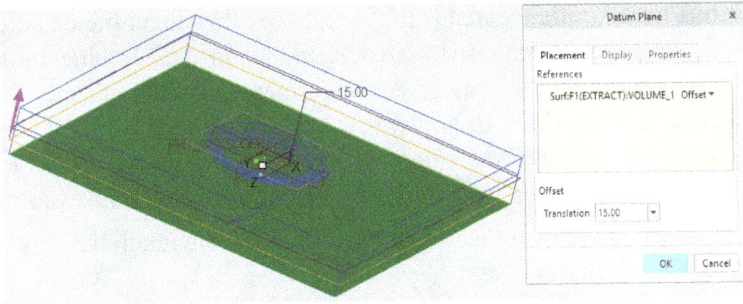

Fig. 8.31.   Creating water lines.

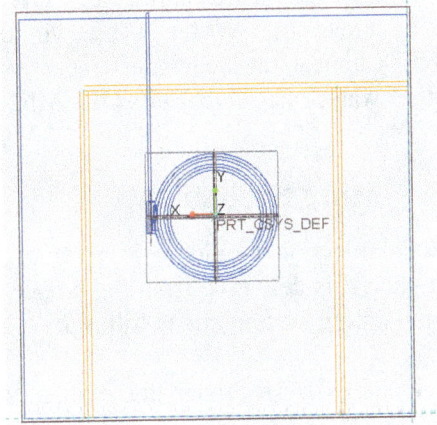

Fig. 8.32.   Water lines.

references. Select the bottom plane of VOLUME_1. Change the direction of the offset to inwards and drag the handle (small white square) of the new datum plane to about **15** mm, a half of the part thickness, as shown in Figure 8.31. Accept with **OK**.

(86) Continue the **Water Line** procedure. In the **SKET VIEW** menu, click on **Default** to confirm. The **Sketch** tool opens. In the **References** window, select the mould coordinate system and click on **Solve**. If the message is **Fully Placed**, then click on **Close**.

(87) Start the **Line Chain** tool (⌣) and draw three lines, the contour of the water line system, as shown in Figure 8.32. The two vertical lines of the contour are, respectively, the inlet and outlet, while the

line in the middle extends to the right to allow an entrance hole for the drilling operation. Rotate the model in the Graphics area to make sure that the contour does not intersect any mould cavities or sliders. The aim is to prevent leakage of coolant (water). For the same reason, one water line system is located in one mould insert only. The second insert might have another separate water line system. If the result is satisfactory, click on the green tick (✔) to close the **Sketch**.

(88) A new window **Intersected Components** opens. Tick the **Automatic update box** to show the names of intersected parts. Ignoring the BLOCK.PRT, there is only one part, VOLUME_1, in the list and no other intersected parts. Click **OK** to close.

(89) Click **OK** to close the **Water Line** window. A new item WATERLINE_1 appears in the Model Tree.

(90) Design another water line system in VOLUME_2 part.

8.14 Adding a Gating System

The gating system in injection moulding is a set of channels through which the molten material is delivered from the nozzle to the cavity(ies). The main parts of the gating system are as follows:

- *Sprue* — A channel that receives the molten material from the nozzle.
- *Runner(s)* — A channel that delivers the molten material from the sprue to the cavity(ies).
- *Gate(s)* — A narrow opening in a mould attached to the cavity though which the molten material is injected into the cavity.

Figure 8.33 shows a four-cavity moulding. The parts are still attached to the gating system. They are supposed to be separated and the gating system recycled.

8.15 Creating a Sprue

The sprue is a channel tapered **3** to **5** degree to facilitate the ejection. The sprue position on the mould depends on the position of the nozzle of the moulding machine. The procedure to create a sprue is as follows:

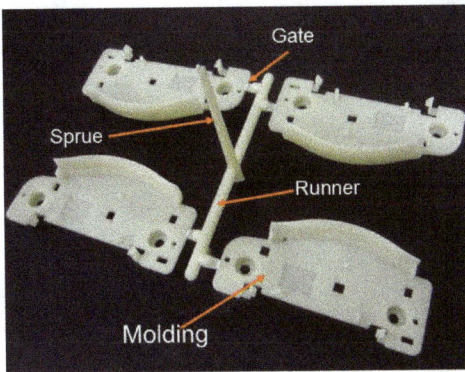

Fig. 8.33.   Example of a four-cavity moulding shot still attached to the gating system.

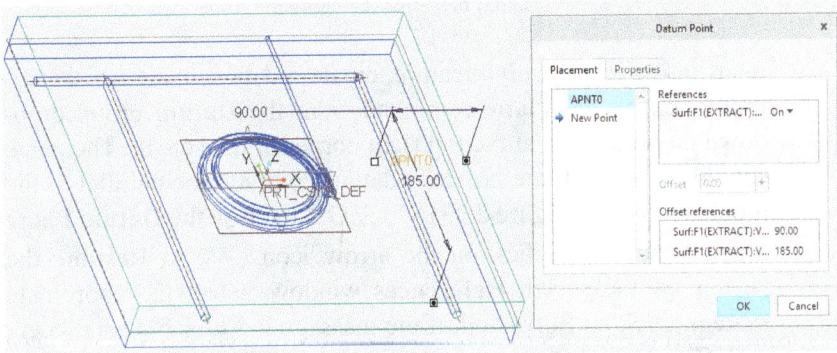

Fig. 8.34.   Datum Point marking the sprue position.

(91) First, create a datum point to mark the position of the sprue. Click on the **Point** icon (⁎⁎), **Datum** group, to activate the **Datum Point** dialogue window. Select the top plane/surface of the mould to create an input in **References**. Click within **Offset references** to activate, press (CTRL + hold) key, and click on the two sidewalls, which are perpendicular to each other (Figure 8.34). Alternatively, pick the two green handles in Graphics area and drag/drop each one onto a sidewall. Enter the point offset values according to the actual nozzle position. Click **OK** to close (**Datum Point** window).

(92) Switch the ribbon to the **Model** tab and select the **Revolve** ( ) from the **Cut & Surface** group (**Model** tab). On the fly and

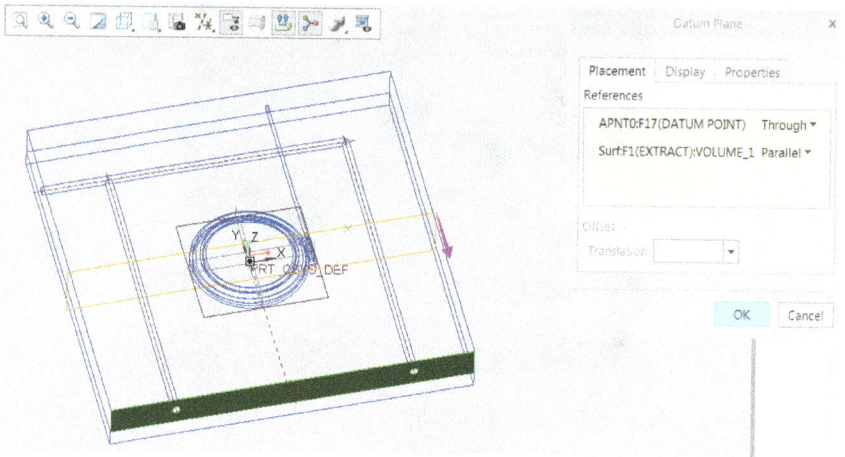

Fig. 8.35.   Datum plane for sketching the sprue.

interrupting the **Revolve** feature, create a datum plane for sketching. Click on the **Plane** command from the **Datum** group dropdown menu located at the top right corner of the screen. The plane should go through the previous datum point and be parallel to the mould side wall (Figure 8.35). Click **OK** to close the **Datum Plane** window and then click on the arrow icon (▶) to **Resume** the sketch. In the sketch **References** window, select the coordinate system, click on **Solve** and, if the massage is **Fully Placed,** accept and **Close**. Change the display style to **Wireframe**. Select the **Centreline** tool (┊) and create a vertical centreline through the created point as the axis of revolution. Use the **Line Chain** tool to sketch a closed contour with the following dimensions: **3** degree taper, **2** mm top side and **4** mm bottom side. Note that this contour starts from the top surface of the mould and penetrates the bottom part by a few millimetres (Figure 8.36).

(93) When the sketch is done, click on the green tick (✔) to close the Sketcher and then close the **Revolve** cut.

8.16 Creating a Runner

The **Runner** command (feature) requires the **Mold Component Catalogue** to be installed by default. If the feature is not present, then run the software SETUP again to install it (Chapter 1, Section 1.7).

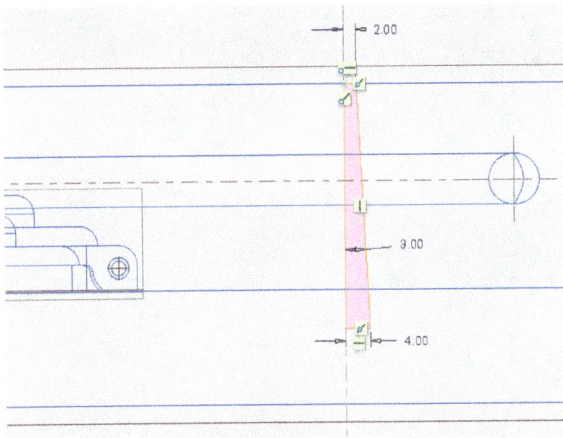

Fig. 8.36.   Sketch section of the sprue to be revolved.

A runner is a channel cut into the mould that starts from the sprue and leads to the cavity(ies) through a gate. Usually, the runners are cut along the <u>parting surface</u> as cylindrical channels. The procedure is as follows:

(94) Change the display style to **Wireframe**.

(95) Click on the **Runner** icon (✳ Runner), **Mold** tab, to activate. In the **Shape** window, select **Round**. In the small window above the mould, enter the runner diameter **5 mm**, and accept with the green tick (✔). You will be prompted to select a sketching plane. Select the <u>parting surface</u> of VOLUME_1 or VOLUME_2 and confirm with **OK**.

💡 Use **Pick From List** if it is not possible to select the right surface. (Click on the surface, press (RMB + Hold), and then select **Pick From List**. This will reveal all geometry entities located at that point. Select the Surf:F1(EXTRACT):VOLUME_1.)

(96) Next, click on **Default** in the **SKET VIEW** window and in the **References** window select the coordinate system as reference for the sketch. In the Sketcher, click on the **Line Chain** (⌒) tool and sketch the runner path. Draw a horizontal line from the centre of the sprue to a point located near (just before) the cavity as shown in Figure 8.37. Continue drawing another horizontal line with an

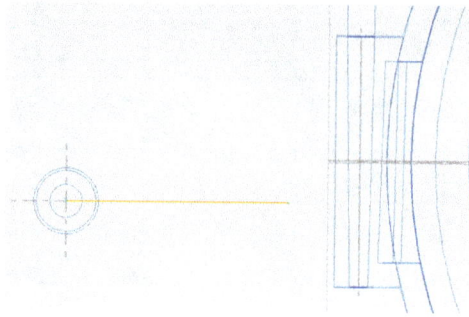

Fig. 8.37.   Drawing the runner's path (1st line).

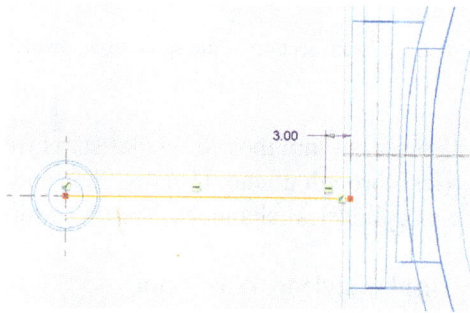

Fig. 8.38.   Dimensioning the gate segment.

endpoint coincident with the cavity vertical edge. The second line is the 'gate'. Use **Dimension** tool (Dimension) and place **3** mm length (Figure 8.38).

Note that the sketch path is just two lines, but in case of multiple cavities it might be a more complicated network of channels.

(97) When finished with the channel path, click on the green tick (✔) in the dashboard. A new window, **Intersected Components**, opens up where the user should identify the intersected volumes. Click on the **Automatic update** to detect the intersecting volumes. VOLUME_1 and VOLUME_2 should appear in the items list. Click on the icon All (▤) to select all and then **OK** to close.

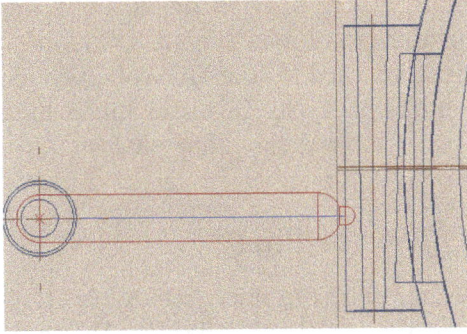

Fig. 8.39.   Segmented runner.

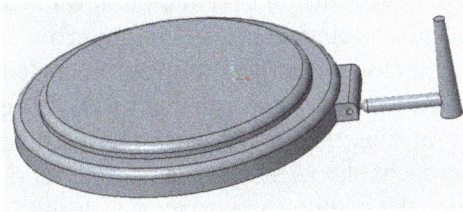

Fig. 8.40.   Moulding attached to the gating system.

(98) Next, assign different diameters to the runner segments. Click on **Segment sizes > Define** in the **Runner** window. The longer segment diameter is set to **5** mm. Select the shorter (gate) segment and accept with **OK**. Change its diameter to **2** mm and click accept (✔) icon.

(99) Close the **Runner Segment** window with **Done/Return** and **Done**.

(100) Select **Preview** in the **Runner** window to view the runner/gate segments (Figure 8.39).

(101) If the result is correct, click **OK** to close.

(102) Open the test shot created earlier. The gating system should be attached to the moulding as shown in Figure 8.40.

Instead of using the segmentation tool, the gate can be created or modified in Part mode to a different shape. Some of the reasons for designing a special gate are to improve the filling, avoid cold weld lines, or create a weak section, which can be easily snapped off to separate the moulding from the gating system, where the latter will go for recycling.

The extension of the sprue cone (Figure 8.40) forms a so-called 'cold slug well', which is supposed to 'trap' the solidified slug formed in the nozzle at the end of the shot cycle. This is an important feature of the gate system design, which facilitates the filling process.

8.17 Creating Ejector Pins

The purpose of ejector pins is to safely remove (eject) the moulding out of the opened mould after plastic solidification. They are standard components selected from a mould catalogue. Each pin (Figure 8.41) has a stem, which slides through a hole in the mould, and a cylindrical head. The head locates the pin in a plate, which moves the pin inwards and outwards. In the case of rectangular components, a minimum of four ejector pins located at the four moulding corners are recommended. In the case of cylindrical components, three equidistant pins at **120 degree** apart are used. The number of pins required to push the moulding out may vary based on the component shape, size and area of ejection. The tip of the ejection pin touches the component surface and leaves visible marks on the finished component. The mould designer should take extra care in the selection of pin locations. An extra ejector pin is used to eject the sprue.

Within the **Mold cavity** module, ejector pins can be created similar to sliders as described earlier in this chapter. However, Creo offers dedicated tools for modelling, including a library of standardised ejector pins. Similar to Section 8.16 the **Mold Component Catalogue** should be installed to use the ejector pin modelling tools. The procedure for creating ejector pins is described in the following steps.

(103) The ejector pins locations are referenced as datum points. Use the **Sketch** tool and create three points on the lid reference part and

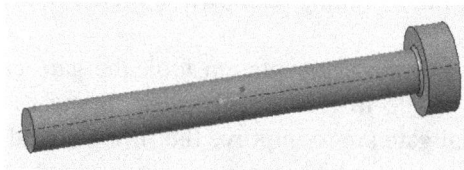

Fig. 8.41.   Ejector pin.

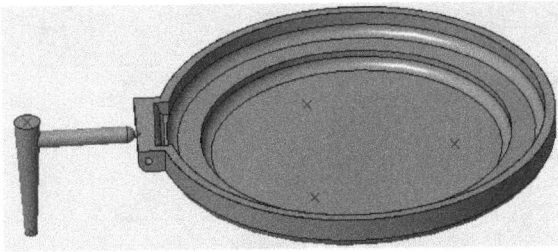

Fig. 8.42. Positions of the ejector pins contact.

then **Point** tool (××) to make a point on the sprue as shown in Figure 8.42.

(104) To create a standard ejector pin, click on the **Catalogue** icon ( ), from the **Mold** tab, **Components** group. Select **Ejector Pin** in the **Menu Manager** window.

(105) Select **Add Set** for several pins or **Add Single** for one. In this case, click on **Add Set** and a new **Define Set** dialogue window opens. Provide inputs as follows: Click on the arrow ( ) and select the four datum points, created earlier (from the Graphics area or Model Tree); Keep the **Set type** as **Identical** for the same pin size.

(106) In the **Component** slot, click on the **Define set members using catalogue** ( ) icon. A new **Define Parameter** window pops up where the user can choose different units, vendors, sizes, etc. In this case, click on the **DIAMETER** down arrow and select diameter **4**, which is the appropriate size in this case. Keep all other preselected parameters unchanged. A drawing and all parameter values of the pin are shown in the window. Type a name for the set, e.g. **Pins** in the **Component Name** slot (at the bottom). Alternatively click on the AB icon (AB↳AB) to generate a name from the catalogue pin number. Close **Define Parameters** window with **OK**.

(107) Now you are back to the **Define Set** window. The **Base plane** slot expects a datum plane that corresponds to the position of the pin head. This is the surface of a separate plate, parallel to the top surface of the mold, where the pin heads are fixed.

(108) This design does not have a separate plate for the pins. A datum plane created 'on the fly' will identify the pins' base plane. Click

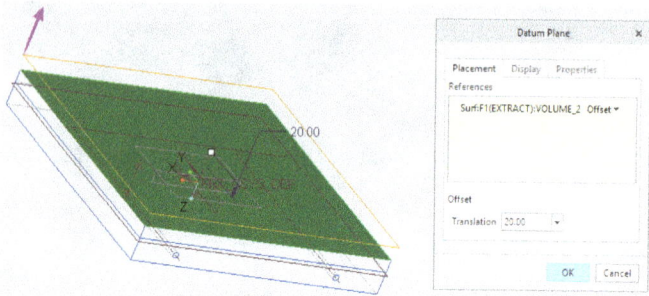

Fig. 8.43.   Ejector pin Base plane creation.

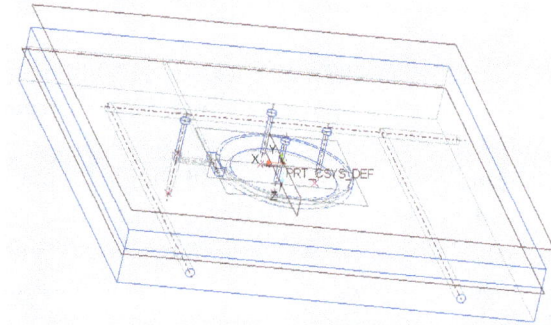

Fig. 8.44.   Mould with the ejector pins.

on the **Plane** tool ($\square$), (**Datum** group, **Mold** ribbon), and create an offset plane from the top surface as shown in Figure 8.43.

(109) Back in the **Define Set** window, click on the arrow ( ) in front of the **Base plane** to activate and select the previous datum plane.

(110) Click on the arrow ( ) in the **Orient plane** and select any side of the mould. Select **Preview** (bottom of the window) to view the current position of the ejector pins. Flip the orientation by selecting the second icon ( ) in the **Base plane** if the pins are upside down (stems pointing outwards). A correct preview is shown in Figure 8.44.

(111) Close the window with **OK** and then click on **Done/Return** in **Component set** window.

An ejector pin set can be redefined or deleted by activating the **Catalogue** ( ) and selecting the **Ejector Pin** command. Notice the **Redefine Set** and **Delete Set** in the COMPONENT SET menu.

(112) The ejector pins' length might not match the distance to the surface area of the moulding to be ejected. The user can modify the length by activating each pin within the assembly and extending or trimming its length to the right distance using the **Extrude** command (add or remove material).

(113) The created ejector pins will interfere with the mould inserts. To correct that, subtract their volume from the mould inserts, as explained in Section 8.12.

(114) Perform interference detection (Chapter 7) and make sure that there are no interfering volumes.

Chapter 9

Computer-Aided Manufacture

9.1 Introduction to the NC Assembly Module

ⓘ The **NC Assembly** is a module within Creo **Manufacturing** that enables the user to create machining tool paths (milling, turning, EDM-ing) and perform virtual reality machining simulation. Various machining strategies can be tested and evaluated in order to select the most efficient one. The user can 'machine' a workpiece and 'manufacture' a 3D model and then process and transfer the information to a real NC machine tool for actual manufacture. **NC Assembly** can determine the machining time for any sequence set, no matter how complicated the tool paths.

Aims:
- To introduce the **NC Assembly** module;
- To understand the basic steps in creating a manufacturing model, environment, operations and simulation.

Outcomes:
At the end of this chapter, the reader will be able to:

- Open the **NC Assembly** module and configure a manufacturing model;
- Create the manufacturing environment: create an assembly, insert an existing reference model (part) and create a workpiece;
- Set up an operation and machine coordinate system for milling sequences;

- Machine a flat surface, generate a profile cut and machine a pocket;
- Create and modify holemaking sequences;
- Prepare and simulate sequences and visually verify the manufacture of a model using **NC Check** simulation;
- Estimate the manufacturing time to cut a part.

9.2 General Workflow and Preparation of the NC Assembly Model

The general workflow for an NC manufacturing modelling is as follows:

(A) Start a new **NC Assembly** model (file);

(B) **Import** a reference model (part);

(C) **Create/assemble** a workpiece (stock) from which the designed part will be machined;

(D) **Define** an **Operation** by selecting a work centre (machine type environment) and creating a machine coordinate system;

(E) **Create** a set of individual **milling sequences** including suitable **cutting tools** and **machining parameters** to machine a part;

(F) **Simulate** 3D visualisations of the machining sequences;

(G) Edit the machining sequences and optimise the machining time;

(H) **Combine** the sequences, observe the simulation of an operation, and generate a cutter location (CL) data output file;

(I) **Post-process** the CL file to create a standard ISO G-code CNC program for use with a specific machine work centre.

There are some important issues that must be considered at the beginning of this lesson, and these are as follows:

- **Files:**
  The main **NC Assembly** model file (*.ASM file) contains the reference model, workpiece, machining sequences (features), and other manufacturing components (for example jigs). In addition, the **NC Assembly** model generates several additional files, for example, a toolpath file. All these files belong to the same manufacturing model and must be located in one directory, i.e. the Working Directory.

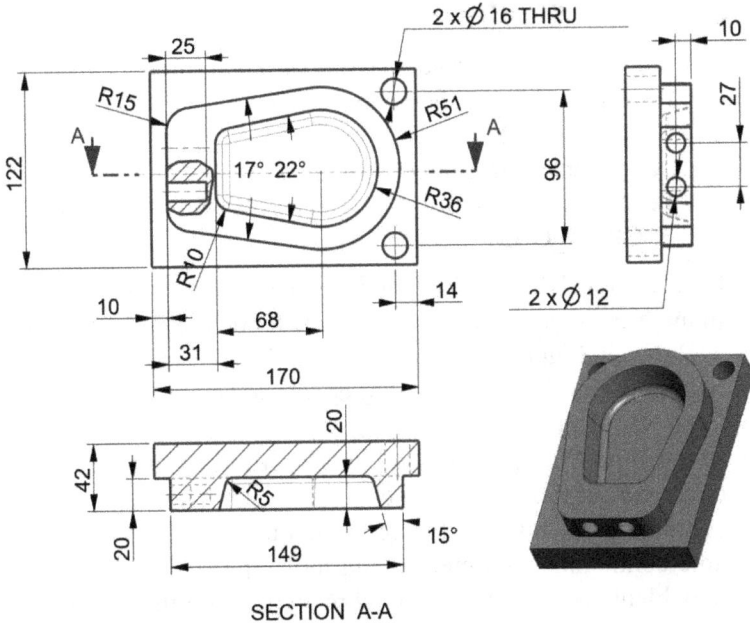

Fig. 9.1.   Drawing of the reference part (model), all dimensions are in millimetre.

- **Working directory:**
  Start Windows File Explorer and create a subdirectory ...\CREO_ CAM in the user directory. Copy the reference model (TESTBASE. PRT, Figure 9.1) into it. After launching Creo, set up the above directory as Working Directory.
- **Preparation for machining:**
  In advance, prepare a logical machining process plan following the actual part geometry and a specific machining strategy. In other words, select a sequence of machining steps that need to be performed to achieve the final reference part geometry, for example, roughing, re-roughing, finishing, pocket milling, hole drilling, reaming, etc. Note, that this tutorial will not advice on how to create the process plan, as this is covered in the manufacturing textbooks.
- **Saving models:**
  Save the models regularly. Sometimes, the software could crash and the user may lose a significant amount of work. Remember that the **NC Assembly** module will not allow to save a model unless all features (or procedures) are closed. Either close or cancel any open feature (operation set up or a sequence) before saving the model.

9.3 Starting the Tutorial

Before starting this tutorial, create the reference part (model) TESTBASE. PRT using the dimensions as shown in Figure 9.1.

(1) Start Creo (unless it is already running).
(2) Set the Working Directory: **File > Select Working Directory** > C:\ USER\CREO_CAM.
(3) Create a new NC assembly model. Click on the **New** icon (⬜) and in the **New** window and choose **Manufacturing** and **NC assembly** as shown in Figure 9.2. Untick the **Use default template** box (at the bottom of the figure) in order to choose a template in Metric units.
(4) Type Name a new name, e.g. CAM, and click **OK**.
(5) In the **New File Option** window, select **mmns_mfg_nc** and then click **OK**. After confirmation of the template, a new window containing the manufacturing modelling interface with all available commands, sub menus, and tools appears (Figure 9.3). The new blank assembly file is now ready to store the manufacturing information.

Important: It is advisable to check what the NC assembly model units are from the start. To do this, go to **File > Prepare > Model Properties**.

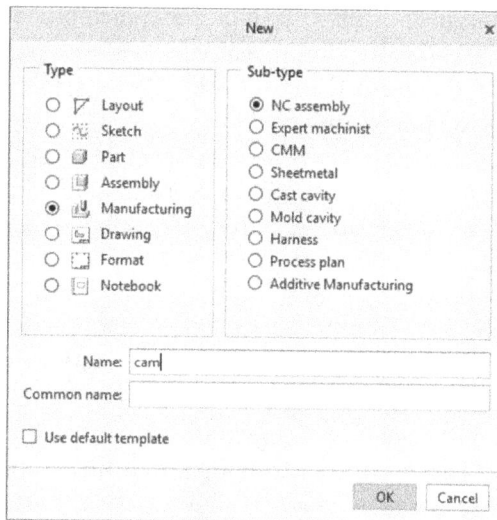

Fig. 9.2.   New file window (Manufacturing type and NC assembly sub-type selected).

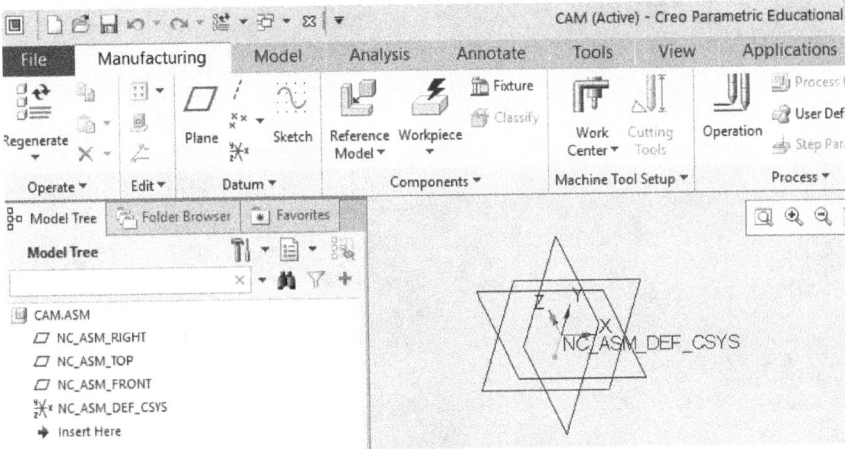

Fig. 9.3. Manufacturing ribbon interface (a fragment).

Find the **Units** row and if the units are **mm** then close the window. If not, click on **Change**, select **millimetre Newton Second** from the **Units Manager** window, tick on **Include submodels** and select **Set**. In the **Changing Model Units** window, keep the option **Convert dimensions**, select **OK** to finish, and click on **Close** (the **Units Manager** window).

(6) Following the general workflow, insert the reference model (or part) into the NC assembly. From the **Manufacturing** tab/**Components** group, click on the **Reference Model** icon ( ). The **Open** window pops up displaying the content of the Working Directory. Select the TESTBASE.PRT and click **Open**. In the **Component Placement** dashboard, select **Default** (from the **Automatic** drop-down menu). The reference model appears as shown in Figure 9.4.

(7) Click on **Save** ( ) to save the model.

9.4 Creating the Workpiece

The workpiece is a part that represents the initial stock material. During the creation of the manufacturing model, the user defines a series of machining 'cuts' or machining sequences in order to remove material from the workpiece and achieve a geometry matching the reference model geometry. The workpiece can be created separately in <u>Part mode</u> and then

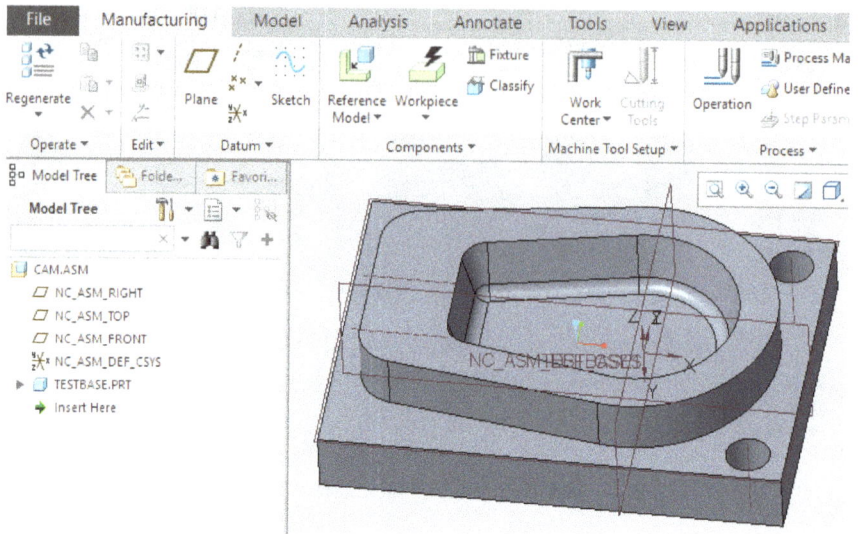

Fig. 9.4.   The imported reference model for machining.

assembled into the NC assembly or directly within the NC assembly model.

(8) If the workpiece was created separately, then it can be assembled as follows: Select the **Workpiece** drop-down menu (Workpiece ▼) icon (not the icon itself (⚡)) from the **Components** group and then click on the **Assemble Workpiece**. If the workpiece shares the same coordinate system with the reference model, quickly align them by changing **Automatic** selector to **Default** alignment. Alternatively, align these manually until it is fully constrained.

(9) It the workpiece is not available as an existing component, then it can be created within the NC assembly model as a solid part using the **Extrude** command as follows: From the **Workpiece** drop-down menu select **Create Workpiece** (⚡). Type the name 'BLOCK' in the **Enter Part name** box, and click the green tick (✔) to accept. The **Menu Manager** window (on the right) will display the feature creation menu. Keep the pre-selected **Solid** option and click on

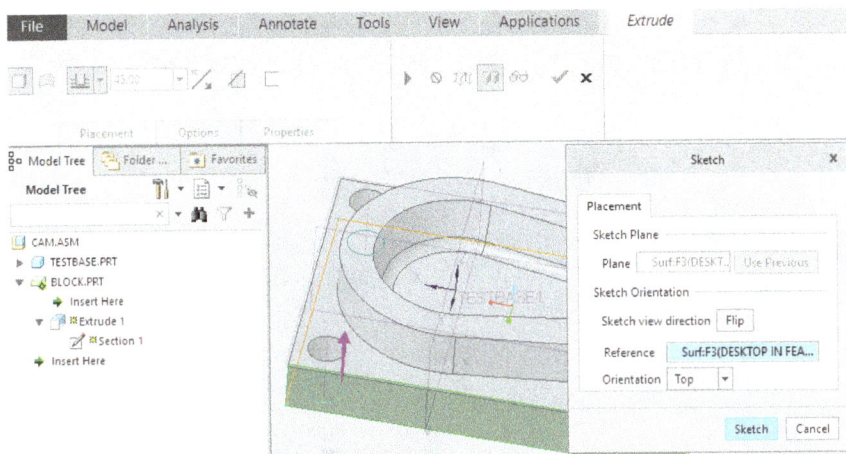

Fig. 9.5.   Sketch direction for the workpiece.

the available option **Protrusion**. In the **SOLID OPTS** window, keep the pre-selected **Extrude** and **Solid** and then click on **Done**. The familiar **Extrude** dashboard appears (the same as in <u>Part mode</u>).

(10) Setting the References. Click on the **Placement > Define**. In this example, select the reference part base surface as sketching **Plane**, and any perpendicular side wall as **Reference**. Note the red arrow denoting the direction for the extrusion. The correct direction should include the reference model as shown in Figure 9.5.

(11) Click on the **Sketch** ( ) icon. If the **References** window shows that there are no available references, select at least two perpendicular surfaces, datum planes or edges from the reference part. Click on **Solve** and if the status is **Fully Placed** (Figure 9.6) then click on **Close** to finish.

(12) Click on the **Project** (Project) tool in the **Sketching** group and select the four edges of the reference part as shown in Figure 9.7 and then click **Close** (in **Type** window).

The **Project** tool will copy the reference part edges and create a section with the same size. If a different size is needed, then use the **Line Chain** tool to draw the section.

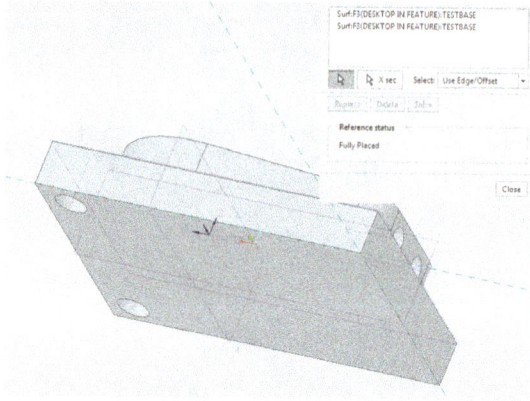

Fig. 9.6.   Sketch references from features.

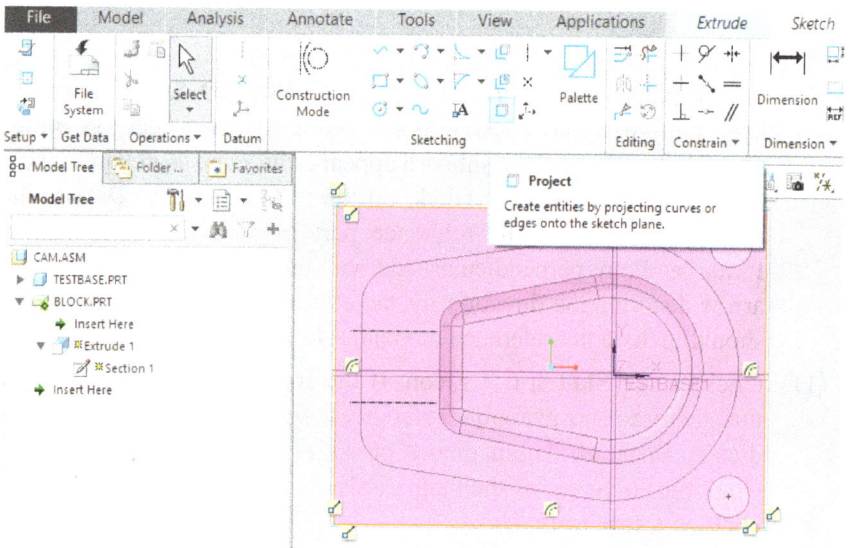

Fig. 9.7.   Projection of existing features.

(13) If the sketched section is pink, indicating a correct closed contour, click on the **OK** icon (✓) to close. Rotate the model slightly (press MMB and drag) in order to view the extrusion direction. Set the extrude depth to **45** mm, i.e. with sufficient allowance for

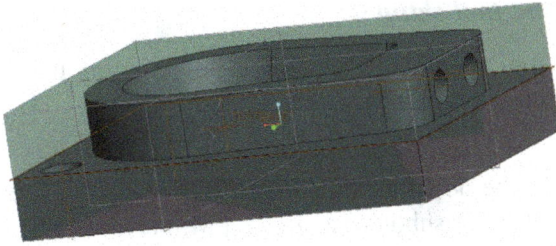

Fig. 9.8.   Completed workpiece extrusion.

machining and click on accept (✔). The workpiece will be displayed in translucent green as shown in Figure 9.8. Notice the assembled parts TESTBASE.PRT and BLOCK.PRT in the Model Tree.

To change the workpiece geometry, click on the arrow before the name of the BLOCK.PRT in the Model Tree to view the features. Click on **Extrude**, to open the quick menu, and select **Edit Definition** ( ).
 (14) Click on **Save** ( ) to save the model.

The saved model will be a manufacturing assembly file with an 'ASM' suffix. After **Save**, the user is prompted to accept the original file name. After launching Creo next time, use **Open** and find the manufacturing assembly (*.ASM) file in the working directory.

9.5 Operation Setup

Operation Setup defines the machine type for the model machining, machine/fixture parameters and machine references as well as retract data. This is mainly technological data associated with a specific machine system, and the user need to collect it in advance or input it later when the specific settings of machining sequences are required.

 (15) Select the **Operation** ( ) icon from the **Process** group, i.e. **Manufacturing** > **Process** > **Operation**. The **Operation** dashboard opens (Figure 9.9).

9.5.1 *Machine tool setup*

The machine tool setup could be done before starting the operation, but often it is sorted out within it.

(16) Click on the down arrow under **Mfg Setup** ( ) icon (top right corner of the dashboard) and select Mill (Figure 9.9).

(17) The **Milling Work Center** dialogue window appears (Figure 9.10). Change the name of the operation and keep the 3 axis unless the plan is to use a 4 or 5 axis machine. Click on **OK** to accept and close the window. Note: Proceed to the next section, but do not close the **Operation** dashboard yet.

9.5.2 *Setting the machine coordinate system*

(18) To set the Machine Coordinate System (programme zero), click on the **Select 1 item** slot (Select 1 item) in Figure 9.9. In this case, there are no suitable machine coordinate systems to select; therefore one needs to be created. Setting a coordinate system has two stages: (a) set the origin position — point, vertices; (b) set the orientation — the alignment of the *X*, *Y* and *Z* axes.

(19) To create a new coordinate system, click on the down arrow in the **Datum** icon ( ), at the top right corner of the screen (Figure 9.9), and select **Coordinate System** ( ).

(a)   Origin position.

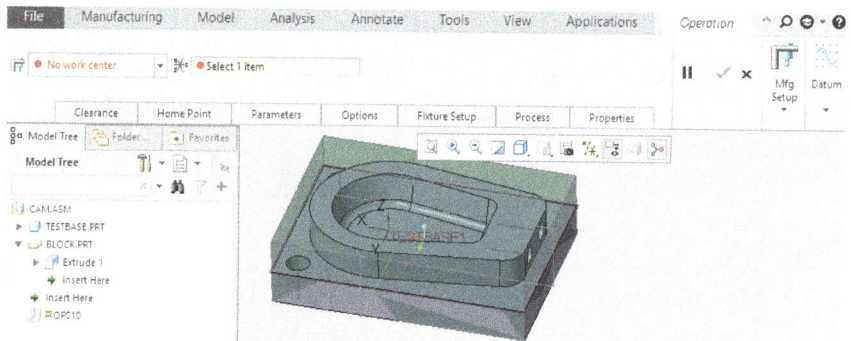

Fig. 9.9.   Operations dashboard.

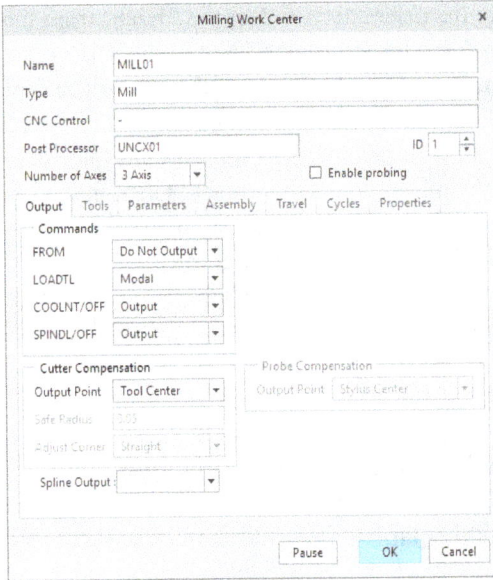

Fig. 9.10.   Milling Work Center window.

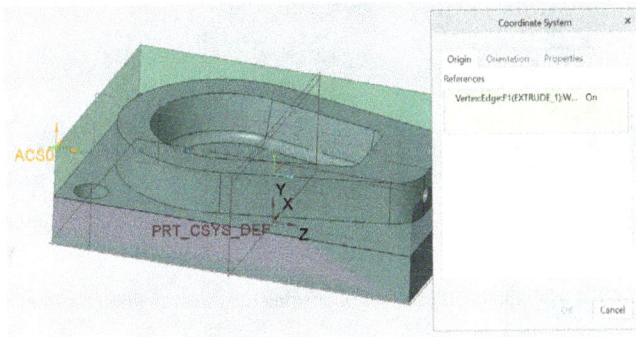

Fig. 9.11.   Selection of the origin.

(20) In the **Coordinate System** window, select the vertex (Figure 9.11) as a coordinate system origin.

(b)   Changing the orientation.

In a 3-axis NC milling, the cutting tool rotation axis, i.e. the tool/ spindle axis, is parallel to the Z axis. The correct orientation of the Z axis is critical.

(21)   To change the orientation, select the **Orientation** tab and then click inside the **Use** slot to activate. Note that X is greyed out in the 'to determine' box. Select an edge of the workpiece for the X alignment (Figure 9.12). The X axis should be aligned to the edge and point to the right. **Flip** to change the direction until correct. Note that X in the 'to determine' box is now black. Click in the lower **Use to project** box and then select an edge on the workpiece to align the Y axis (Figure 9.13). Note that the alignment of the Z axis has been set automatically. Manipulate the edge selection and possibly flip direction until the orientation setup is correct as shown in Figure 9.14.

The **Orientation** of the basic machine coordinate system is now complete. Note that the same workpiece can have several machine coordinate systems each having a different orientation. On a 3-axis milling machine, each coordinate system may represent a different workpiece set up (orientation on the machine table).

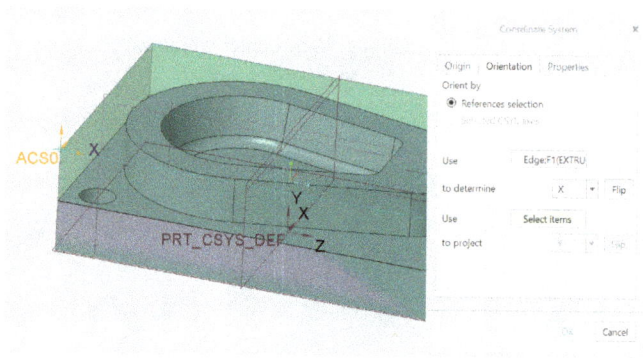

Fig. 9.12.   Orientation selection (Coordinate System dialogue window).

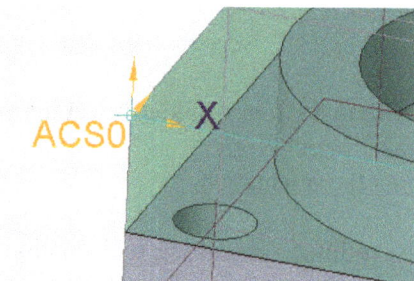

Fig. 9.13.   Selection of the X axis edge.

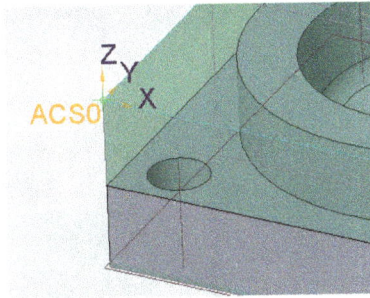

Fig. 9.14. *X*- and *Y*-axes correct directions.

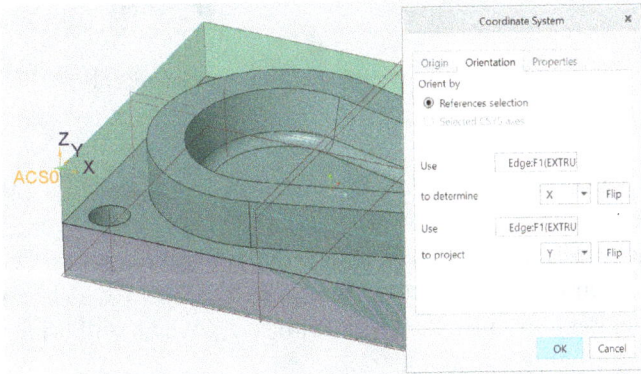

Fig. 9.15. Final Coordinate System setup.

(22) When the location and orientation of the machine coordinate system are correct, click **OK** to close the **Coordinate System** window (Figure 9.15). Note that a new item **ACS0** (coordinate system) has been added to the Model Tree. Click on the **Resume** icon (▶) to activate the **Operation** dashboard.

(23) Notice the red dot in the coordinate system ( ) slot. Click inside (to add item) and then select **ACS0** from the Model Tree as an input.

9.5.3 *Setting up the retract surface (clearance)*

(24) Click on the **Clearance** menu in the **Operation** dashboard. Change the **Type** to **Plane**. Click in the **Reference** slot to activate and select the top surface of the workpiece. In the **Value** slot, specify the height

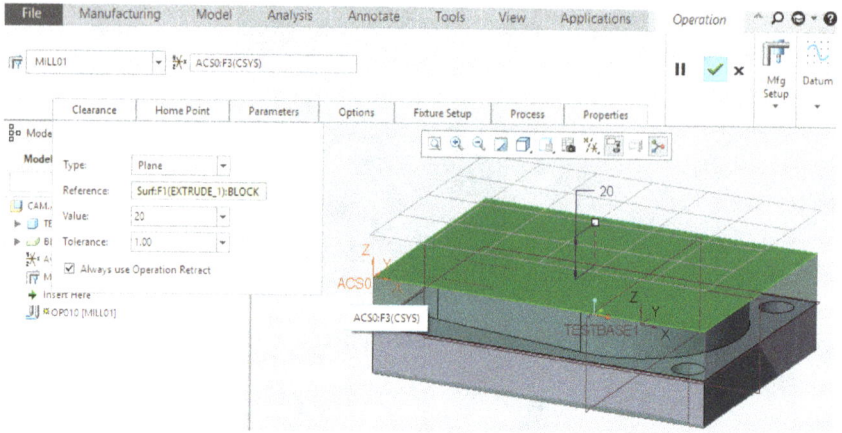

Fig. 9.16.   Retract clearance (Operation dashboard).

of the retract surface from the selected surface, e.g. type **20** mm and press ENTER (Figure 9.16). Keep the default tolerance and click the **Clearance** tab again to complete the setup of retract surface.

(25) If all inputs are provided, the green tick (✔) will be available. Click on it to finish the **Operation** setup, which will bring back the **Manufacturing** ribbon.

9.6 Creating Machine Sequences

Prerequisites in the creation of manufacturing data within the CAM module are as follows:

- Selection of a suitable machine — mill, lathe, mill-turn or Wire EDM.
- Cutting tools selection (tool material, type, size). These tools will be used in the actual machining sequences.
- Calculation of the cutting tools parameters.

In addition, the spindle rpm (speed) and feed rates to be applied to each tool and workpiece material should be calculated in advance. This data is critical because the **NC Assembly** uses this information for the determination of the sequence time.

The tool geometry (tool size) information and machinability data given throughout this tutorial is for guidance purposes only. The users must examine their visual and data outputs and make changes where necessary in order to perform the tasks correctly.

The sequences developed for this tutorial have not been optimised. The user can reflect and consider other possible toolpath strategies and improvements of the process plan.

In an actual manufacturing environment, there are additional important considerations concerned with the tool holding, jigs, fixtures and clamping of the workpiece. These considerations are beyond the scope of this book and will not be included in the tutorial text.

9.6.1 *Selecting the machine sequences*

(26) Select the **Mill** tab. The **Milling** and **Holemaking** groups appear in the ribbon. Note the variety of sequences available in the **Milling** group, including those listed under the pull-down menu. It is advisable to select the most appropriate sequence type that corresponds to the manufacturing operation from the process plan.

(i) The most frequently used sequences and their applications are as follows:

- **Face milling** — Usually used for machining flat faces. The tool (typically an end mill cutter) moves with feeding along X and Y at a certain constant depth in Z.
- **Profile milling** — Similar to face milling, but the tool (end mill or side mill) cuts around a 3D feature with vertical or tapered (roughing) walls.
- **Pocketing** — Similar to the above two, but used to machine cavities (pockets) with vertical or tapered (roughing) walls and a flat bottom.
- **Surface milling** — One of the most powerful type of sequences. Cuts with feeding along all three axes X, Y and Z. Usually used for finishing very complex (often free-form) shapes using a ball nose cutter.
- **Engraving** — Used to cut small shallow cavities with a small size tool.
- **Volume/window milling** — Usually used for roughing i.e. quickly removing the bulk of the material from a particular volume or within a specified-in-advance window.

ⓘ For drilling (**holemaking**), which can be carried out on a milling machine, the sequence list is as follows:

- **Standard drilling** — It is used for making blind or through holes.
- **Countersink** — It is usually used with a special tool (countersink) to machine chamfers, to make a conical hole, or to mark the position for drilling a hole without pre-existing starting hole.
- **Face** — It creates facing holes or machining counterbores features. The tool is usually an end mill cutter or special counterbore tool.
- **Deep drilling** (also known as peck drilling) — It is used for drilling deep holes with depth to diameter ratio of 5:1 or larger. In this type of sequence, the depth of the hole is divided into sections. The drill plunges to the end of each section, and then retracts out of the hole in order to clear off the swarf and prevent clogging and/or breaking.
- **Break-chip drilling** — It is similar to deep drilling, but without coming out of the hole. This option is usually chosen if vibrations of the tool (long thin drill bit) are expected and when multiple retracts and plunges in the same hole might cause tool or hole damage.
- **Boring** — It is usually used to enlarge previously drilled holes by means of a boring cutting tool.
- **Tapping** — It is used for cutting threaded holes.
- **Reaming** — It is used for hole finishing and achieving high accuracy.

The selection of appropriate sequences and their ordering is a process planning activity. The process plan for machining of the part shown in Figure 9.1 could be as follows:

(a) Face milling the top face; Tool: TC (Tungsten Carbide) face mill **150** mm diameter;
(b) Profile milling the top shape; Tool: TC end mill **50** mm diameter;
(c) Drilling **16** mm diameter holes; Tool: HSS (High Speed Steel) drill **16** mm diameter;
(d) Roughing the tapered pocket; Tool: TC ball end slot drill **12** mm diameter;
(e) Finishing the tapered pocket; Tool: TC ball end slot drill **10** mm diameter;
(f) Reorient the part;
(g) Drilling **12** mm diameter holes; Tool: HSS drill **12** mm diameter.

9.7 Face Milling

The sequence types described earlier have very similar procedures of application, although each one has some specifics in its menu. This section will provide general principles in the applying of the face milling sequence. This is the first step from the process plan.

(27) Select the **Mill** tab (unless already selected).

(28) Select the **Face** milling (⊥) icon to open its dashboard. Click on the **Tool Manager** icon (T) at the top left on the **Face** Milling dashboard. The **Tools Setup** dialogue window opens (Figure 9.17).

(29) Specify the tool dimensions in the **Geometry** area. Provide inputs and fill in all slots, i.e. **Name**, **Type**, **Material**, **Units** and **Number of Flutes**. Following the previous entries, fill in the following data: **ENDMILL150**, **END MILL**, **HSS**, **Millimetre** and **4** (Figure 9.17).

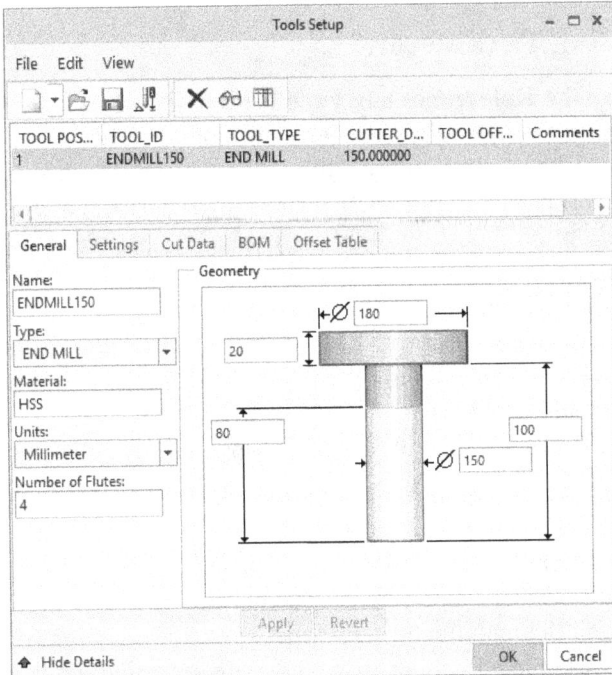

Fig. 9.17.   Tools Setup window.

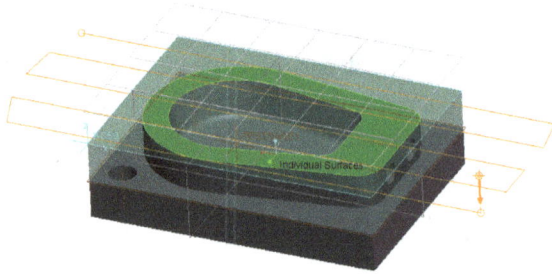

Fig. 9.18.   Surface selection for face milling.

The tool dimensions are **180** (top diameter), **150** (tool diameter), **80** (tool cutting height), **100** (tool height) and **20** (top height) — all in millimetres. To finish, click on **Apply**. The first tool will appear as position **1** in the **Tools Setup** window.

(30) Close the **Tools Setup** window with **OK**.

9.8 Cutting Parameters

(31) Select the **References** tab, **Face Milling** dashboard.
(32) Click in the **Machining References** box to activate.
(33) Select the top <u>surface of the reference model</u> (not the top of the workpiece) (Figure 9.18). Use (RMB + Hold) and **Pick From List** option (Section 4.3.9, Chapter 4) to make the correct selection.
(34) Select the **Parameters** tab (highlighted in yellow). In the dialogue window (Figure 9.19), the yellow colour is a 'flag' indicating that a value must be entered in that slot. This window shows the Basic parameters that are generally used in any sequence.
(35) For a full selection of machining parameters, click on the icon (⬚) at the bottom right corner of the window (Figure 9.19).

✏️ The **Cut Feed**, also known as feed rate, and the **Spindle Speed** need to be calculated. Use appropriate machining data and corresponding formulae from the manufacturing textbooks. After rounding off and converting the calculated values into the corresponding units, enter the

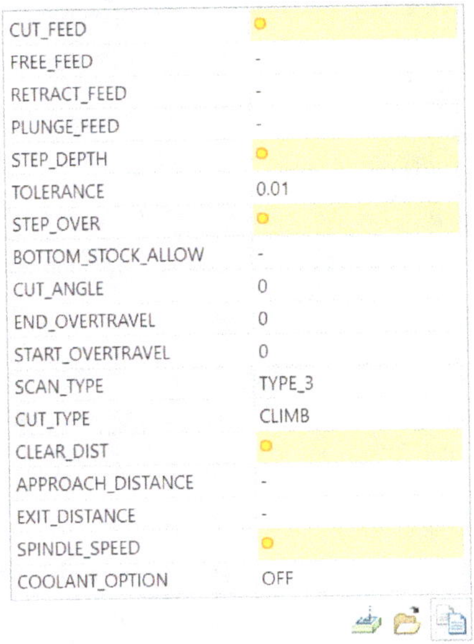

CUT_FEED	○
FREE_FEED	-
RETRACT_FEED	-
PLUNGE_FEED	-
STEP_DEPTH	○
TOLERANCE	0.01
STEP_OVER	○
BOTTOM_STOCK_ALLOW	-
CUT_ANGLE	0
END_OVERTRAVEL	0
START_OVERTRAVEL	0
SCAN_TYPE	TYPE_3
CUT_TYPE	CLIMB
CLEAR_DIST	○
APPROACH_DISTANCE	-
EXIT_DISTANCE	-
SPINDLE_SPEED	○
COOLANT_OPTION	OFF

Fig. 9.19. Milling parameter window.

feed rate in <u>mm/min (millimetres per minute)</u> and spindle speed in <u>RPM (revolutions per minute)</u>.

There are some 'cut type' parameters that require different settings such as **step depth** (dependent on the cutter type) and **step over** (depending on the cutter diameter and on the sequence type).

(36) To enable the tool to retract at a clearing distance, use a positive value in the slot **CLEAR_DIST**.

Enter appropriate values for your example in all yellow slots of the **Parameters** panel. In this example, use the following:

CUT_FEED = 100,
STEP_DEPTH = 5,
STEP_OVER = 30,
CLEAR_DIST = 2,
SPINDLE_SPEED = 1000.

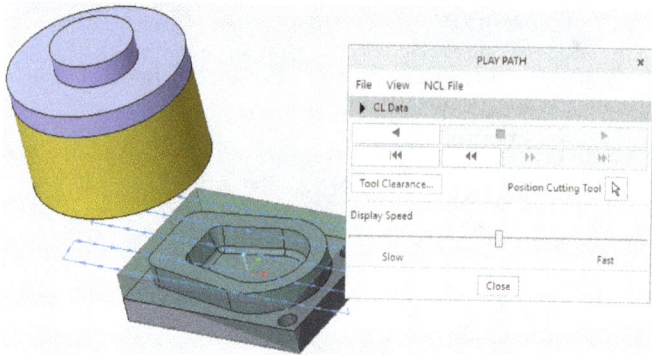

Fig. 9.20.   Toolpath simulation (Play Path).

(37) Click on the **Parameters** tab again to close it. If the **Reference** and **Parameters** tabs are still highlighted, it means that some parameters have been missed and need corresponding inputs.

(38) Review the resulting toolpath by clicking on the ( ) icon.

(39) Click on the **Display Toolpath** icon ( ) to view the tool traversing the tool path. In **PLAY PATH** window, click on **Play Forward** ( ) icon. Notice that the tool (Figure 9.20) follows the outline of the selected model surface. For better control, use the **Display Speed** slider for fast/slow rewind.

(40) Click on **Close** in the **Play Path** window when finished.

(41) Click on the green tick ( ) to apply and save the sequence.

(42) Click on **Save** ( ) to save the model.

9.9   General Procedures to Add, Edit or Play a Sequence

9.9.1 *Add/edit a sequence*

To add a sequence from the **Mill** tab, select a sequence and follow the procedure as described above.

To edit a sequence, select it with the LMB from the Model Tree and from the mini menu click on **Edit Definition**. (The RMB click on a sequence/feature opens the full feature editing menu.)

9.9.2 *Change/add a tool*

In order to change/add a tool, click on **Manufacturing** tab and then on the **Cutting Tools** (⌦) icon from **Machine Tool Setup** group. The **Tools Setup** window opens.

Note that each tool must have a <u>unique</u> number in **TOOL POSITION** column. Therefore, when adding more tools, select the **Settings** tab and set a **Tool Number**, which represents a physical position in the tool changer, different for each tool. Click on the **Apply** icon to add the tool to the list.

There are several types of simulations within Creo Manufacturing, and two of them are as follows:

* The simplest simulation type is using the **Play Path** and trace the toolpath. This simulation can also run from within the sequence or from the Model Tree in a similar way as described in the previous section.
* The second simulation is very useful as it is a virtual reality **Material Removal Simulation**. This requires the simulation software called NC Check to be enabled.

(43) To enable the **NC Check**: Click on **File** > **Options** > **Configuration Editor** (Figure 9.21). Click on **Find**, in the **Type** keyword box type **nccheck**, and then click on **Find Now**. Once **nccheck_type** variable is found, use the **Set value** dropdown arrow to change the value from default (Figure 9.22) to **nccheck**, and then click on **Add/change**. **Close** the **Find Option** window, and **OK** to close the **Creo Parametric Options** window. Accept to save this option in the CONFIG.PRO configuration file by **Yes** and **OK**.

Once activated, the **NC Check** runs as described in the following section.

9.10 Simulating Material Removal

(44) To simulate material removal with the **NC Check**, select a sequence from the Model Tree, and then click on **Material Removal Simulation** (⌦) icon from the mini menu. This opens the **Menu Manager**. Click on **Run** to simulate.

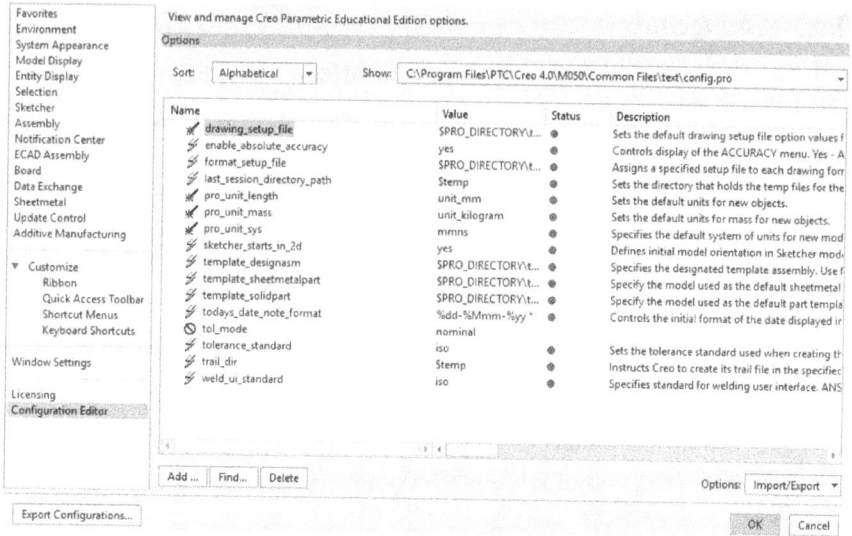

Fig. 9.21.   Configuration Editor options.

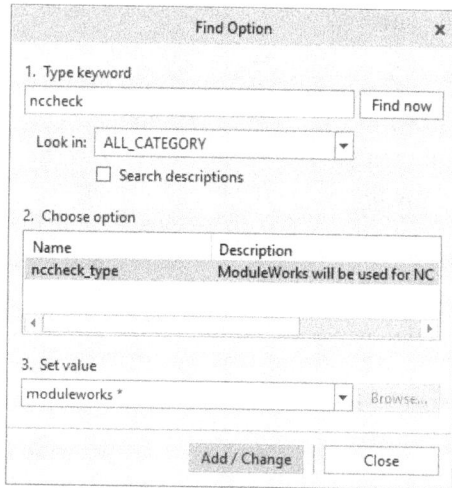

Fig. 9.22.   Find Option window (Configuration Editor).

(45) If the simulation runs too fast, make an adjustment by selecting **Step Size** from the **NC DISP** menu and click on **Enter**. In the next menu, select **Enter** again. In the new window **Enter step size**, which will open under the dashboard, type a smaller step size e.g. **3** and click

on the green tick (✔). Click on **Run** again to resume. This will slow down the simulation for a more detailed observation.

(46) Click on **Refresh** and use the (MMB + drag) to rotate the model for a better view, and then **Run** again. A view of the **NC Check** output screen for the **Face Milling** sequence is shown in Figure 9.23.

The **NC Check** displays the model in the following colours:

Green = the colour of the workpiece;

Navy blue = the colour of the workpiece where the tool is cutting during the sequence and/or some material remains to be cut;

Purple = the colour of the reference model where the tool touches the surface, i.e. all material has been cut away;

Cyan/light blue (not shown in Figure 9.23) = Area where the tool may be grazing or gouging (overcutting) the reference model. The tool/tool path has to be altered to avoid that.

(47) Click on **Done/Return** to exit the **Menu Manager**.

9.11 Sequence Information

The manufacturing model, i.e. the NC assembly file contains the technical information that has been entered in all sequences.

(48) To access and check the correctness of the data, select the sequence in the Model Tree with RMB click, and then select **Information** > **Feature Information**.

This opens the **Feature info** window. A large amount of information is available in several sections on this window. Scroll to the bottom and find the **Manufacturing Info** section. Scroll down within this section and find all details including the tool and sequence parameters and machining time. Use this information for the cutting processes optimisation.

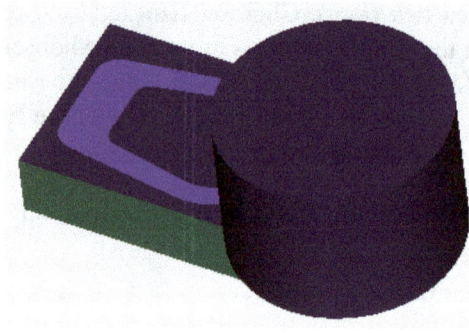

Fig. 9.23.   Face milling simulation (NC Check).

Click on the icon (✖) (top right corner) to exit.

(49) Click on **Save** (💾) to save the model.

9.12 Profile Milling

ⓘ The **Profile Milling** sequence is intended to cut a rounded or angular outer profile of a part. It has a similar dashboard and required inputs as the **Face Milling** sequence.

(50) To create a profile milling sequence, click on the **Mill** tab and select 🔲 Profile Milling. The **Profile Milling** dashboard opens. Note the defaults: the tool and coordinate system. Creo assumed that the previous tool will be used, which might not be the case.

9.12.1 *Profile milling tool*

(51) If a different tool is required, then click on the **Tool** icon 🔧 to open the **Tools Setup** window. In this case, use an <u>end mill</u> as a cutting tool with **4** flutes and **50** mm diameter. Create a new tool providing a relevant name (ENDMILL50) and dimensions. In the same window, go to **Settings** tab/menu and assign the new tool a different/ unique **Tool number** (e.g. 2) to avoid overwriting the previous one

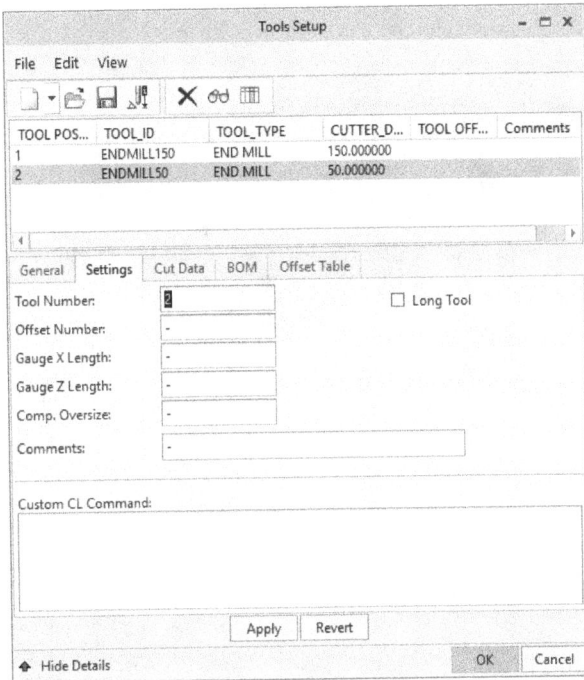

Fig. 9.24.   Tools Setup — providing different tool numbers.

(see Figure 9.24). It helps to remember this by thinking that every new tool will need a different position in the machine tool changer.

(52) Click on **Apply** to add the new tool to the list and then **OK** to close the **Tools Setup** window.

9.12.2 *Selecting the surfaces for profile milling*

Note that in the **Profile Milling** sequence dashboard the two tab menus (**Reference, Parameters**) are highlighted, which means that a corresponding input is required.

(53) Select the **Reference** tab and click inside the **Machining References** slot to activate. Select all external profile surfaces of the reference model (not from the workpiece) as shown in Figure 9.25. Use **Pick From List** option to select the right surface (Section 4.3.9, Chapter 4) if required. Press (CTRL + Hold) key for multiple

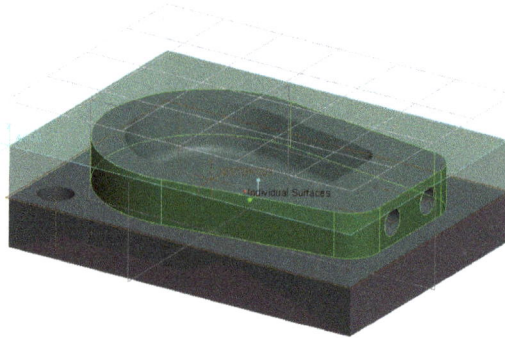

Fig. 9.25.   Selected surfaces for profile milling.

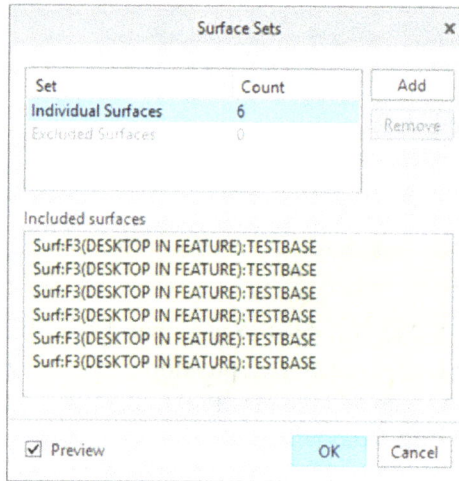

Fig. 9.26.   Surface Sets.

selection. If a wrong surface is accidentally selected, then click on
Details box (under the **Machining References**). In the additional
Surface Sets window, review and edit the list of selected surfaces
as shown in Figure 9.26.

(54) Click **OK** to close the **Surface Sets** window (if it is still opened).

(55) Now click on the **Parameters** tab, which is still highlighted, and
set appropriate values in all highlighted boxes in a similar way as
in Face Milling section. For example, use feed rate of 120 mm per

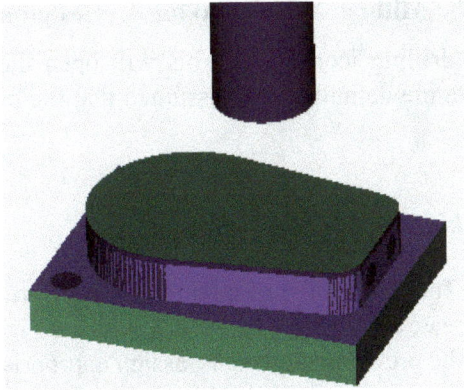

Fig. 9.27.   NC Check for Profile Milling 1 sequence.

minute, spindle speed **900** rpm, etc. When done, click on **Parameters** again to close it. Run **Play Path** ( ) simulation to preview the sequence.

(56) Click on accept (✔) icon to close the sequence.

(57) Select the **Profile Milling 1** sequence/feature in the Model Tree and click on Material Removal ( ) simulation (Figure 9.27).

(58) Review the simulation and check that the sequence performs as expected.

(59) Click on **Save** to save the model. It is important to save the work at this stage. If there are some errors, these can be corrected later using **Edit Definition**.

Note that each sequence simulation runs independently in stand-alone mode assuming that there are no previous sequences, i.e. no Face Milling sequence in this example, that remove material from the work-piece. How to process an ordered set of sequences will be shown later in this tutorial.

9.13 Hole Making Sequences

A sequence from **Holemaking Cycles** group (**Mill** tab) will be used to drill **16** mm diameter standard holes.

(60) Click on the **Mill** tab > **Holemaking Cycles** group and select the **Standard** drilling icon ( ). This will open the **Drilling** dashboard. Note the defaults. Creo assumed that the previous tool will be used.

9.13.1 *Holemaking tools setup window*

(61) Select the **Tool** icon ( ) to open the **Tools Setup** dialogue window and create a suitable tool for drilling (see Figure 9.28).

(62) Similar to the previous sequences, assign appropriate values for the **Name**, **Type**, **Material**, **Units** and sequence parameters. Change the **Type** to **DRILLING**. The **Geometry** settings are specific to a standard drill bit. Make sure that the tool dimensions such as diameter, length, etc. are adequate for the workpiece size. Enter the dimensions shown in Figure 9.28. Starting from the top: **20** mm

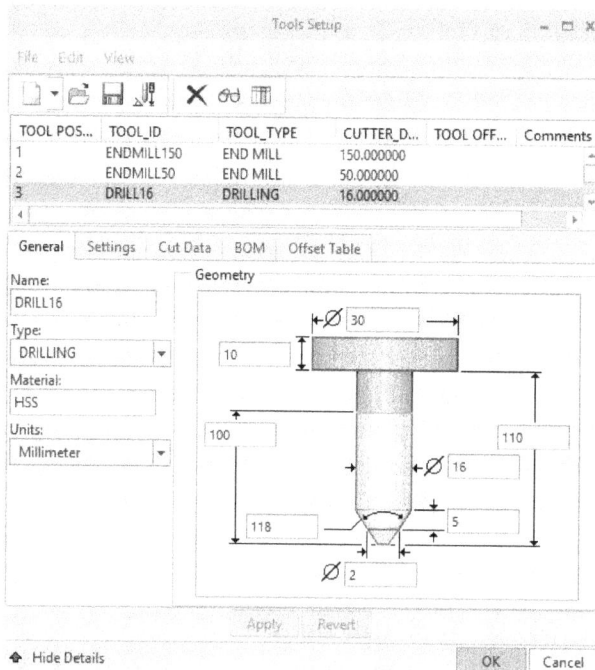

Fig. 9.28.   Tools Setup window.

shank diameter, **10** mm shank height, **110** mm tool height, **100** mm effective drill height, **16** mm drill diameter, **118** degrees angle of the cone, **5** mm cone height, and **2** mm nose diameter.

(63) Do not forget to go to the **Settings** tab and assign a new tool number **3.** This will avoid overlapping with another tool position.

(64) Click on **Apply** to add the tool to the list and **then OK** to close the **Tools Setup** window.

9.13.2 *Holemaking sequence parameters setup*

(65) First, select an appropriate coordinate system with Z axis along the hole axis. Click on the coordinate system slot in the dashboard to activate and then select a coordinate system.

(66) Click on the highlighted **References** tab to define the set of holes to be drilled and the drilling depth.

(67) Click in the **Holes** box to activate the selection. Press (CTRL + Hold) and then click (LMB) on the two axes of the vertical holes. To modify the selection, click on **Details...** and then in **Holes** window (Figure 9.29) edit the selection.

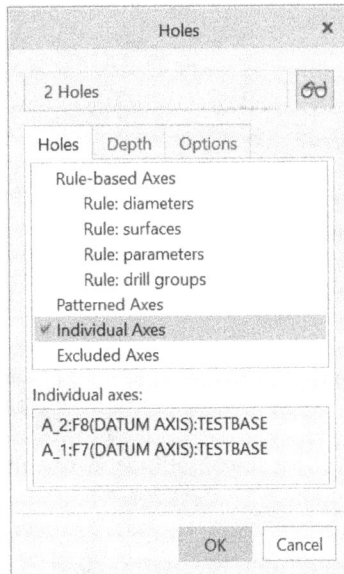

Fig. 9.29. Holes window.

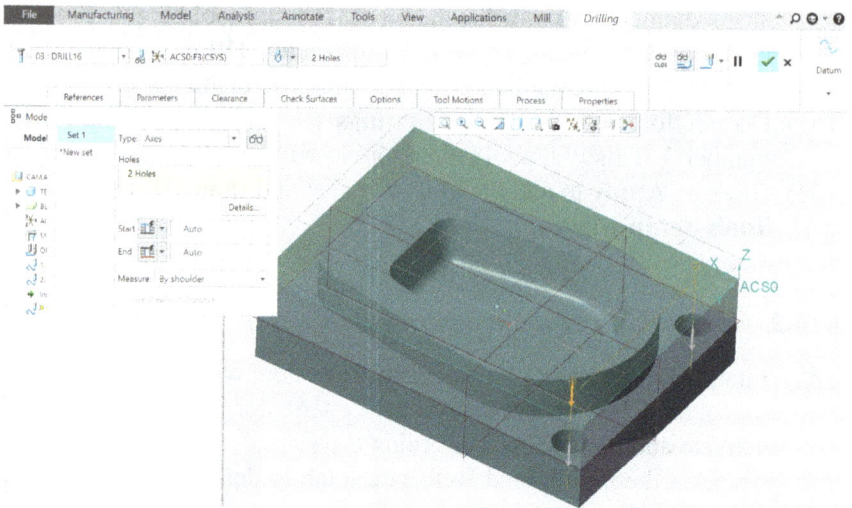

Fig. 9.30.   The Drilling dashboard — holes positions.

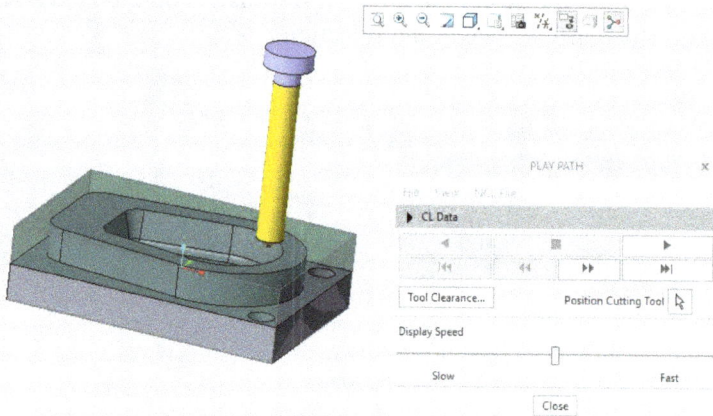

Fig. 9.31.   Play Path simulation — Drilling (hole) sequence.

(68) Select **Accept** (✓) to close and return to the **Drilling** dashboard (Figure 9.30).

(69) Click on the **Parameters** tab, which is still highlighted, to input the drilling parameters (highlighted). When defining these parameters,

consider the cutting tool material: High-Speed Steel (HSS) or Tungsten Carbide. The machinability data is different for each material. For the HSS material, the data is as follows: the surface speed is about **30** m/min, and feed per tooth is **0.2** millimetres. This will provide (needs to be calculated) a **SPINDLE_SPEED** of **600** rpm and feed rate (**CUT_FEED**) of **240** mm per minute. **CLEAR_DISTANCE** is **2** mm.

(70) Click on the **Parameters** tab again to close the window.

(71) Click now on the **Play Path** ( ) icon to run the simulation as described before (Figure 9.31). Also, click on ( ) icon to run the NC Check to simulate the sequence with material removal.

(72) Click on the green tick ( ) to accept and save the sequence. Note that the Model Tree updates with the new sequence.

(73) Click on **File** > **Save** ( ) to save the model.

9.14 Milling (Roughing) Tapered Cavity

This sequence is intended to rough cut the tapered cavity of the part. **Surface Milling** sequence is one of the several options (sequences) that could be used for this task. Suppose that the following cutting tool has been selected: HSS ball mill cutter (ball nose cutter), **12** mm diameter (**2** teeth). The machining data is as follows: spindle speed = **3600** rpm and feed rate = **720** mm per minute.

Continue the previous model and create a new NC sequence as follows:

(74) In the **Mill** tab click on the **Surface Milling** icon (Surface Milling). This will open the **Menu Manager** window (Figure 9.32). The default setups are indicated with tick marks. These include the **Tool** (to allow the setup of an additional tool), **Parameters**, **Surfaces** and **Define Cut** setups. Each setup opens a dialogue window or a sub-menu with corresponding inputs. These are explained in the next sections.

(75) Select **Done** to accept and move to the first setup window.

Fig. 9.32.   Menu Manager.

9.14.1 *Tools setup window*

Following the list of selected setups in the **Menu Manager**, Creo will open the first one — the **Tools Setup**. Use the same procedure as in the previous sections to input required tool parameters.

(76) For roughing, select either a **BULL MILL** or **BALL MILL** type cutter. Use the values shown in Figure 9.33 for this exercise. Change the **Tool number** to another number, e.g. **4**. If the tool number is duplicated, then correct the number. In the warning message window that will appear (with a red circle), click on **Move**.

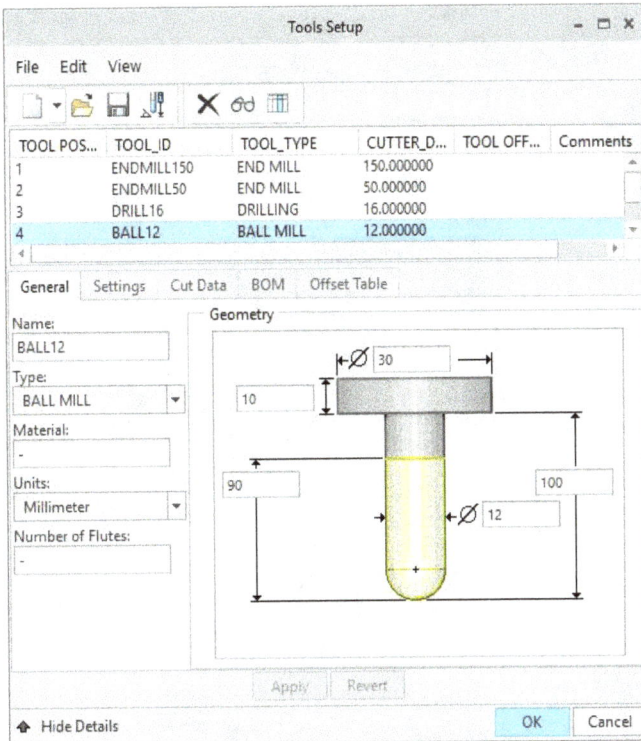

Fig. 9.33.   Tools Setup window.

(77) Click on **Apply** to add the tool to the list and **click OK** to close the **Tools Setup** window and to go to the next dialogue window.

9.14.2 *Sequence parameters setup*

(78) The next dialogue window, **Edit Parameters of Sequence,** opens. The user should enter the sequence parameters as shown in Figure 9.34. Set up all highlighted parameters for **Surface Milling** such as **CUT_FEED — 720, SPINDLE_SPEED — 3600** and **PROF_STOCK_ALLOW — 1** mm. The **1** mm allowance will leave some material for a subsequent finishing. Without this

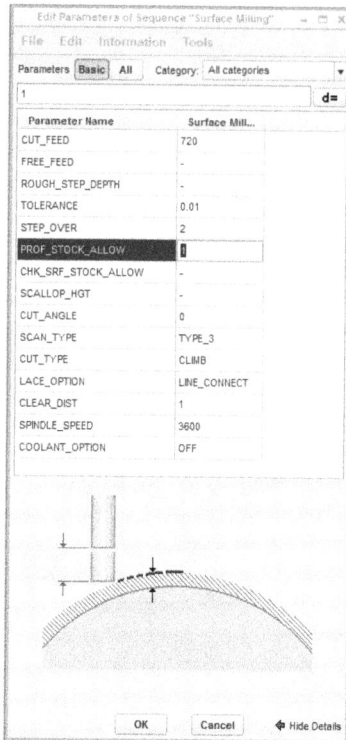

Fig. 9.34.   Edit Parameters window.

allowance, the sequence will cut away all the material and there will be none left for a finishing sequence.

(79)   Click **OK** to close the window.

9.14.3 *Surf pick window*

This setup option will bring up the **SURF PICK** sub-menu (of the **Menu Manager**) for surface selection (Figure 9.35).

(80)   Keep the pre-selected **Model** and click on **Done**, which opens the selection window. Press (CTRL + Hold) and select <u>all surfaces of the cavity including the floor</u>, **13** surfaces in total (Figure 9.36).

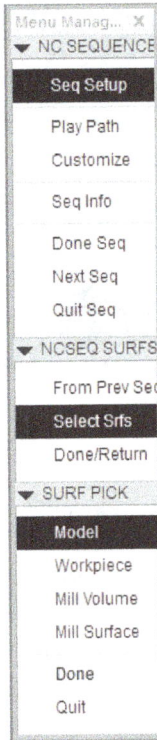

Fig. 9.35.   SURF PICK sub-menu.

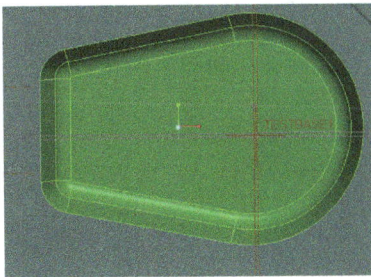

Fig. 9.36.   The selected surfaces.

Click **OK** to accept the selection, click on **Done/Return** to accept again, and then **Done/Return** to close the last window.

(81) Click **OK** to accept the **Cut Definition**.

Fig. 9.37.   NC Check of surface milling.

(82) Run the sequence simulation and observe the results. From the **Menu Manager** window, click on **Play Path** menu and run the **Screen Play** (Play Path) and/or NC Check simulations with material removal. The NC Check screen should look similar to what is shown in Figure 9.37.

(83) If the simulation is acceptable, then click on **Done Seq**.

(84) Click on **Save** (⊞) to save the model.

9.15 Finishing the Tapered Cavity

The next task of the process plan is finishing the tapered pocket. The **Pocket Milling** is one of several available sequences. The radius of the pocket rounds is **5** mm. Therefore, the cutting tool could be a ball mill, with a **10** mm or smaller diameter. The sequence parameters can be similar to the previous sequence. However, to achieve a better finishing quality, the cut depth and step over should be smaller. Feasible machining parameters are as follows: spindle speed = **4500** rpm and feed rate = **900** mm per minute.

(85) To create a pocket milling sequence, select the **Mill** tab and click on the **Milling** group arrow (Milling ▾) to open the drop-down menu with various options/icons as shown in Figure 9.38.

(86) Select **Pocketing**. The **Menu Manager** window opens. Notice that the **Tool** option is not selected (ticked). This is because the system

CAM (Active) - Creo Parametric

Applications | Mill

Roughing
- Face
- Re-rough
- Profile Milling

- Surface Milling
- Finishing
- Corner Finishing

Trajectory Milling

- Engraving
- Custom Trajectory

- Round
- Chamfer

Milling

- Manual Cycle
- Auxiliary

- Thread Milling
- Plunge Rough
- Pocketing
- Swarf Milling

- Pencil Tracing
- Local Milling

- Trajectory
- Cut Line Milling
- Rest Finish

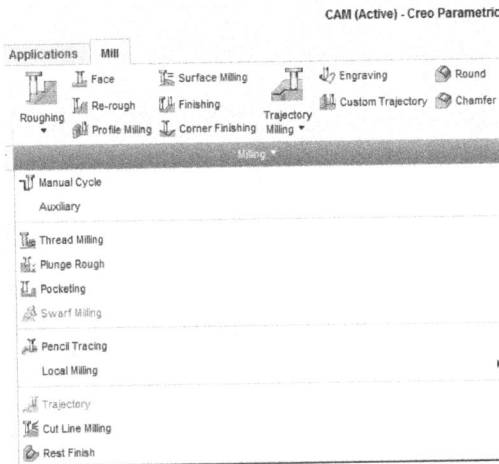

Fig. 9.38.   The Milling group containing various cutting sequences.

assumes that the last tool will be used again. Select (tick) the **Tool** and click **Done** to accept. Similar to Section 9.14, each selected setup, starting with the **Tools Setup** (Figure 9.39), needs appropriate entry values corresponding to the specific process sequence. Do not forget to change the tool number in **Settings** to a higher value, i.e. **5**. Click on **Apply** to add the tool to the list and **OK** to close the window.

9.15.1 *Sequence parameters setup*

(87) The previous step will open the **Edit Parameters of Sequence** dialogue window (Figure 9.40). Provide all highlighted cutting parameters with inputs, as shown in Figure 9.40. Note that **PROF_ STOCK_ALLOW** is set to zero as this sequence is aiming to finish off the surfaces. When done, click **OK**.

(88) The **SURF PICK** sub-menu opens. Keep the pre-selected **Model** and click **Done**, to bring up surface selection window. Press (CTRL + Hold) and select all the surfaces of the cavity including the floor. Click **OK** to accept. Click on **Done/Return** to close.

(89) Now click on **Play Path** and run the **Screen Play** and/or NC Check simulations with material removal (Figure 9.41).

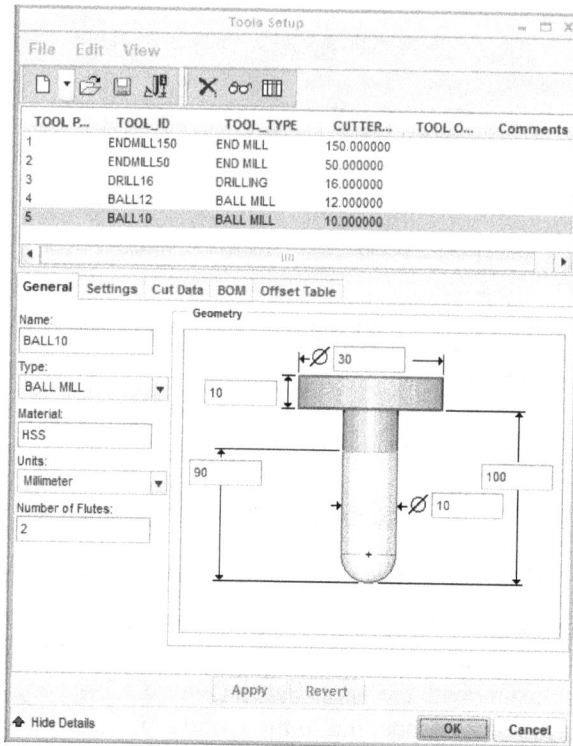

Fig. 9.39.   Tools Setup for pocketing ball mill tool.

Notice the lighter colour, patchy due to the low resolution of the simulation, indicating that no more material allowance is left.

(90) If the simulation is satisfactory, click on **Done Seq**.

(91) Click on **Save** (▥) to save the model.

9.16 Another Hole Making Sequence

This sequence will drill the two **12** mm horizontal holes on the part side. The selected tool is a drill made of HSS material and is of **12** mm diameter. The spindle speed is **800** rpm and the feed rate is **320** mm per minute.

Note that this sequence will require a new coordinate system with the **Z** axis in horizontal direction (or normal to the part side).

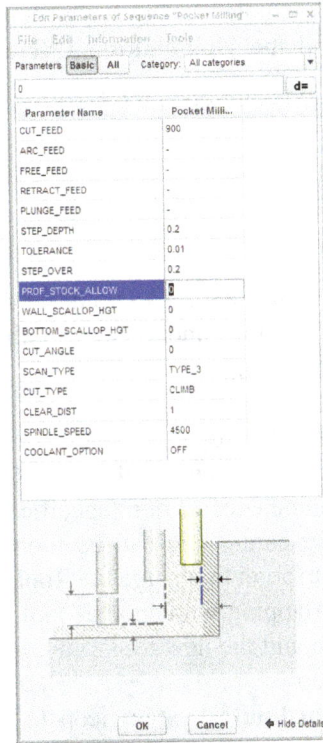

Fig. 9.40.   PROF_STOCK_ALLOW set to zero for finishing.

Fig. 9.41.   NC Check finishing simulation.

(92) To create a standard hole making sequence (or **Drilling**), select the **Mill** tab and find the **Holemaking Cycles** group. Click on the **Standard** ( ) icon and open the **Drilling** dashboard.

9.16.1 *Tools setup*

By default, the system assumes that the last tool will be used in this sequence. Click on the **Tool** icon ( ) (from the dashboard) to open the **Tools Setup** dialogue window and create another tool following the instructions from Section 9.13.

(93) Create a new tool with properties as follows: the **Name** is **Drill12**, **Material** is **HSS** and **Type** is **DRILLING**. The **Geometry settings** include a diameter of **12** mm (specific for a standard drill bit), and the remaining parameters can be found in any drilling tools catalogue. Do not forget to change the **Tool Number** to **6** (**Setting** tab) to avoid overlapping with another tool.

(94) Click on Apply to add the new tool to the list and **OK** to close the **Tools Setup** window.

(95) Click on the **Coordinate System** icon (✳), in **Datum** group, to create a new coordinate system (as shown in Section 9.5.2). Set the Z axis parallel to the side holes axes of and pointing to the part outside (Figure 9.42).

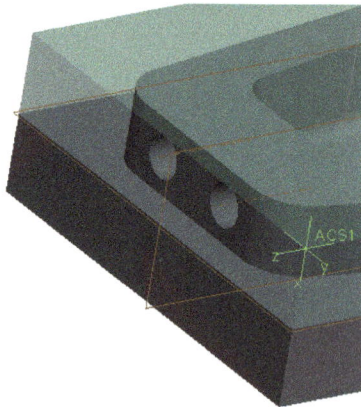

Fig. 9.42.   CSYS for drilling.

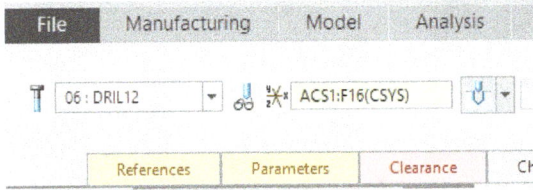

Fig. 9.43.   Selecting new coordinates.

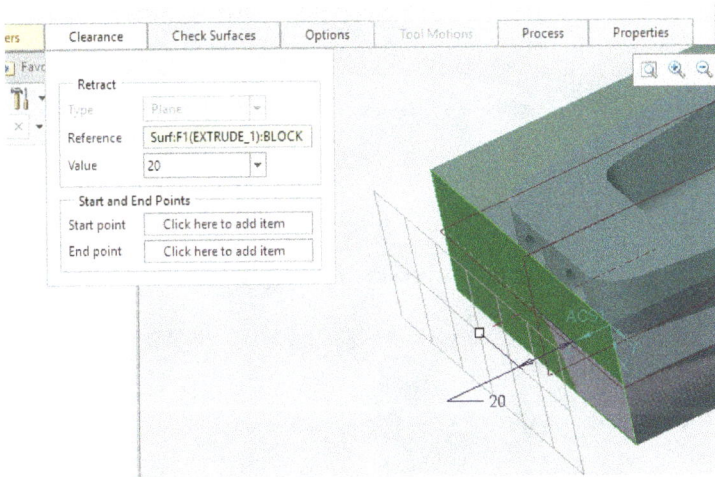

Fig. 9.44.   Setting up of a new Clearance plane.

(96) Click in the **Coordinate System** slot in the **Drilling** dashboard to activate and replace the current with the newly created coordinate system **ACS1** (Figure 9.43).

Notice that the **Clearance** tab is now highlighted in red to indicate that due to the new orientation, the Retract surface requires a different reference plane normal to the Z axis. Sort this out as explained in Section 9.5.3 and shown in Figure 9.44.

Provide the required entries in the **Parameters** tab similar to the previous drilling sequence (Section 9.13.2).

9.16.2 *Setting the hole depth*

(97) Click on **References** tab and follow the same procedure that was used earlier in drilling. Select the holes axes (Figure 9.45).

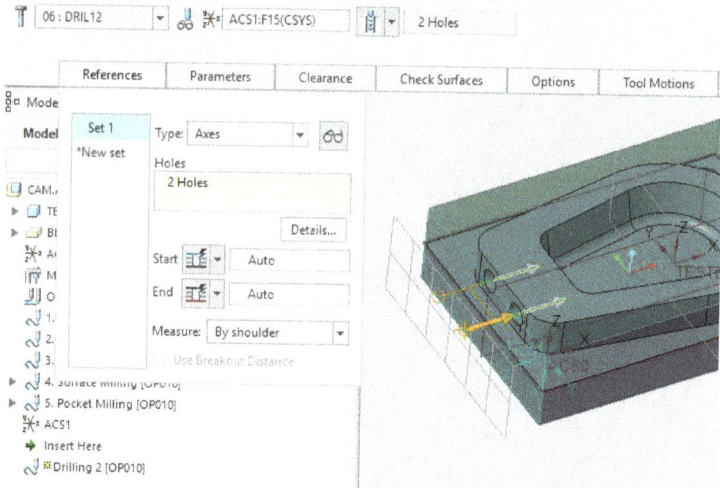

Fig. 9.45.   Hole references selection.

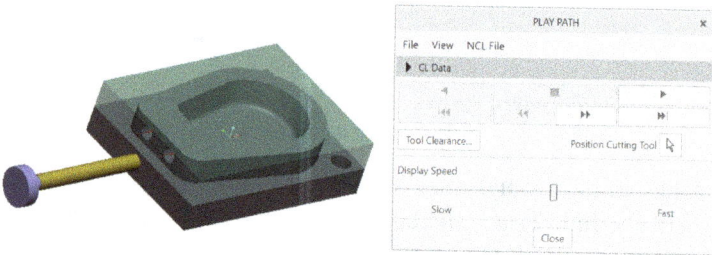

Fig. 9.46.   Play Path of the Drilling sequence.

Keep the default **Start** and **End**, that specifying the hole depth, options set to **Auto**. Close the **References** panel.

(98) Run **Play Path** and/or NC Check as previously described. The **Play Path** screen view will be similar to that seen in Figure 9.46.

(99) If the simulation is acceptable, click the green tick (✔) to save the drilling sequence.

(100) Click on **Save** (🖫) to save the model.

9.16.3 *Drilling deep holes*

When drilling deep holes, with depth more than five times the hole diameter, there is a danger of breaking the tool inside the hole. In this case, it is recommended to use **drilling with pecking** option. It enables the drill bit to retract outside the hole in order to remove the swarf every time of drilling a certain small depth.

(101) To enable that, select the newly created drilling sequence and click on **Edit Definition** to open the **Drilling** sequence dashboard. Click on the pull-down arrow ( ) and change the option from **Standard** to **Deep drilling** ( ). Immediately, the **Parameters** tab will be highlighted. This is because a new parameter **Peck Depth** has been introduced and set to **5** mm depth. The user can input another value if needed.

(102) Run **Play Path** and notice the 'pecking' effect.

(103) The third option available to facilitate the drilling process is under the same drop-down menu ( ) and is called **Break-chip drilling** ( ). It works in a similar way as the **Deep drilling**, but the drill retracts just a small depth away, enough to break the chip without leaving the hole.

9.17 Machining Time and Optimisation

In order to optimise the machining time, the user needs to identify the sequences with the longest times. A convenient way to find information about this is to use the **Process Manager** ( ) tool located in the **Manufacturing** tab, the Process group.

(104) In the new **Manufacturing Process Table** window, by default the cutting time is not displayed. To reveal it, click on the **View Builder** icon ( ), then in **Not Displayed** section click on the down arrow, and change to **Mfg Info Parameters**. From the list,

Fig. 9.47.   Manufacturing Process Table showing the Machining time.

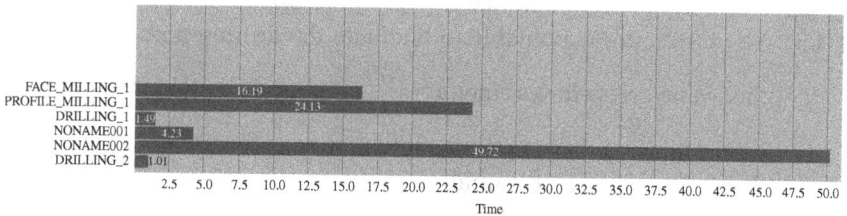

Fig. 9.48.   Process Graph.

select the **Machining Time (Min.)** parameter and click on the arrows (>>) icon to transfer it to the **Displayed** column. Click **OK** to activate. Scroll to the right and the **Machining Time (Min.)** will be displayed in the last column, as shown in Figure 9.47.

To get a maximum effect in machining time reduction, focus on the longest sequence(s) from the table in Figure 9.47. A graphical representation of these sequences can be activated by clicking on **Process Graph** (⊞), Figure 9.48. If necessary, use the zoom tool to see the sequence names on the *Y*-axis. In this case, the longest sequence is **NONAME002-Pocket Milling-finishing**.

(105) Close the graph and table and try to figure out a way to speed up the process by changing some of the sequence parameters.

(106) Click on that sequence in the Model Tree and select the **Edit Step Parameters** (⬱) icon. In the **Edit Parameters of Sequence**, change some of the parameters to speed up the process.

Another way to reduce the process time is to improve the <u>Process Plan</u> by introducing new and/or replacing some sequences with other that could be faster. For example, including roughing sequences with larger tool feed rates or sequences with larger diameter tools. Remember that this is only a virtual simulation. The change of parameters should be based on realistic forecasts specific to the actual machine, available tools and manufacturing capable to achieve faster rates for the selected material.

✎ When making changes by replacing a sequence, instead of deleting the sequence, use **Suppress** command (RMB click on the sequence in the Model Tree and then click on **Suppress** from the mini menu) to temporarily exclude the sequence from the regeneration process (calculation).

9.18 Operations Simulation

9.18.1 *Reordering of sequences (features)*

ⓘ Often, an already created sequence has to be moved up or down the Model Tree list to satisfy certain technological reasons or correct some errors. This can be done by selecting the sequence (feature) in the Model Tree and then dragging and dropping it. Click (LMB + Hold) on a sequence and slowly drag and drop this sequence to a new position in the Model Tree. It will be successful only if there is no violation of the "Parent–Child" relationship involved between the selected feature and the affected features (see Chapter 2). In other words, when a sequence refers to a "parent" feature and the user tries to reposition this feature before its "parent" then the "parent" feature will be highlighted in blue and the system will not allow the move.

After arranging all sequences in their correct order in the Model Tree, they can be joined into a single cutter location (CL) file. This file could be then used to simulate the entire operation.

9.18.2 *Joining all sequences*

(107)   Go to the **Manufacturing** tab > **Output** group, click on **Save a CL file** (CL File ▾) down arrow, and from the dropdown menu select **Save a CL file**. **Menu Manager** window opens. Select **Operation** and click on the operation **OP010** to bring up the **PATH** menu. Select **File** and in the **OUTPUT TYPE** menu keep the defaults selected (ticked). Click on **Done** to bring up the **Save a Copy** window.

(108)   In the **File Name** slot, enter a name, or keep the default **op010**, for the Cutter Location (CL) data file and then click **OK**. A CL data file (*.NCL type) is generated from the cutter paths of all sequences within the operation. Click on **Done Output** in the **PATH** menu (**Menu Manager**) to exit.

9.18.3 *Play path simulations for the complete set of sequences*

(109)   The next step is to simulate the whole operation for the sequence set. Click on the **Play Path** (Play Path ▾) down arrow, **Manufacturing** tab > Validate group, and notice the two familiar simulations available: **Play Path** and **Material Removal Simulation**.

(110)   Click on **Play Path,** select the *.NCL file created earlier, i.e. op010.ncl and then click **Open** to run it. The **DISPLAY CL** menu opens. Click on **Done** to run and observe the simulation (Figure 9.49).

(111)   In a similar way, the user can run the **Material Removal Simulation** (NC Check) for the same operation (Figure 9.50).

(112)   Click on **Done/Return** to exit the simulation.

If some modifications have been done to the sequences, the CL data file (*.NCL file) created earlier will not update automatically. The user needs to generate a new CL data file for the modified sequence set and run the simulations.

9.19 Post Processing

The CL data file (*.NCL file) has to be converted into a CNC machine-specific programme that can be executed. To do that, the user needs to

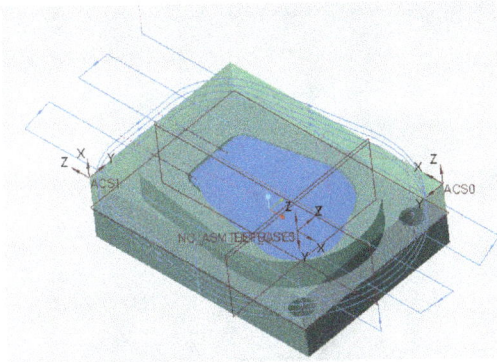

Fig. 9.49.   Play Path simulation.

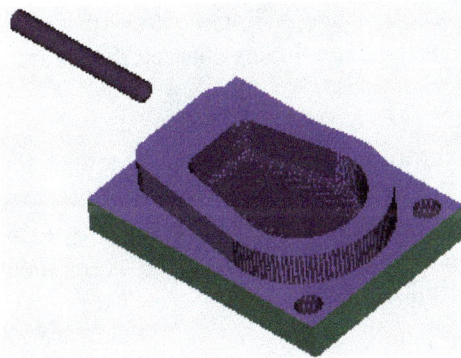

Fig. 9.50.   NC Check simulation.

post-process the *.NCL file to standard ISO G-code commands as follows:

(113) Click on the **Post a CL file** (Post a CL File) icon, **Manufacturing** tab, **Output** group, select the *.NCL file, i.e. OP010 and click on **Open**.

(114) Keep the defaults in the **PP Options** window and click **Done**. From the **PP LIST** (post-processors), select one corresponding to the actual CNC machine controller, for instance **UNCX01.P12** for the FANUC controller. A black window will flash and the **INFORMATION WINDOW** will open displaying the log file (Figure 9.51).

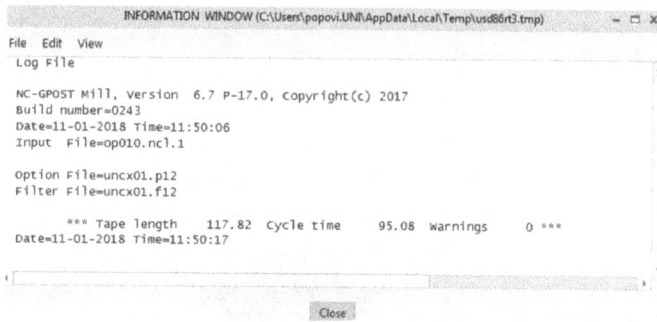

Fig. 9.51.   Information Window.

If the post-processing fails to show the **INFORMATION WINDOW**, then check whether folder name (or path) of the Working Directory is too long or if it contains characters that are not allowed (like spaces).

(115) Close the information window and examine the CNC program text. Find this text file in the Working Directory with the name NAME_OF_OPERATION.TAP. Open it by using any text editor such as Windows WordPad or Notepad. A fragment of this long file is shown in Figure 9.52.

9.20 Additional Machining Operations

Often, due to a variety of manufacturing constraints, the machining of a part cannot be done only by milling. **NC Assembly** offers additional machining operations such as the following:

- Lathe (turning) — Applicable for rotational parts with central symmetry, for example shafts;
- Mill-Turn — A combination between turning and milling machine;
- Wire Electro Discharge Machining (EDM) — Also known as spark erosion. It is used to cut through holes with vertical or tapered walls.

There is a similarity between turning and milling, and the reader can use this tutorial for creating a turning operation. However, the wire EDM is a different process that will be shown in the next section.

```
op010.tap - Notepad
File  Edit  Format  View  Help
N5  G71
N10  ( / CAM)
N15   G0 G17 G99
N20   G90 G94
N25   G0 G49
N30  T1 M06
N35  S1000 M03
N40  G0 G43 Z20.  H1
N45  X230.885 Y10.81
N50  Z1.
N55  G1 Z-3.  F100.
N60  X-53.796
N65  X-61.859 Y36.097
N70  X234.986
N75  X235. Y61.394
N80  X-64.433
N85  X-61.859 Y86.691
N90  X234.986
N95  X230.885 Y111.979
N100  X-53.796
N105  Z20.
N110   G0 G49
N115  T2 M06
N120  S2000 M03
N125  G0 G43 Z20.  H2
N130  X72.127 Y136.188
N135  Z-2.
N140  G1 Z-8.  F100.
N145  X150.803 Y124.653
N150  G2 X185. Y85.077 I-5.803 J-39.576
N155  G1 Y37.712
```

Fig. 9.52.   G-code programme opened in Windows Notepad.

9.20.1 *Wire EDM*

In some cases, milling or drilling cannot be applied, and so an alternative manufacturing process has to be introduced. For example, when machining vertical through holes and in cases where the workpiece is made of hardened steel, conventional drilling is not feasible. In this case, wire EDM is a much more efficient option.

(116) Continue the model from the previous sections. Instead of drilling the two **16** mm vertical holes, a new operation (different from milling) will be added to the model to cut these holes.

(117) Go to the **Manufacturing** tab and click on the **Operation** ( ) icon. Select the **Mfg Setup** icon ( ), open the pull-down menu, click on the **Mfg Setup** down arrow and select **Wire EDM** (Figure 9.53).

(118) The **WEDM Work Center** dialogue window opens as shown in Figure 9.54. Change the name of the operation and keep the 2 axis unless the plan is to use a 4 axis Wire EDM machine.

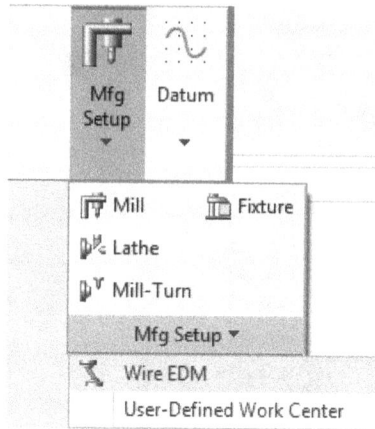

Fig. 9.53.   Mfg Setup.

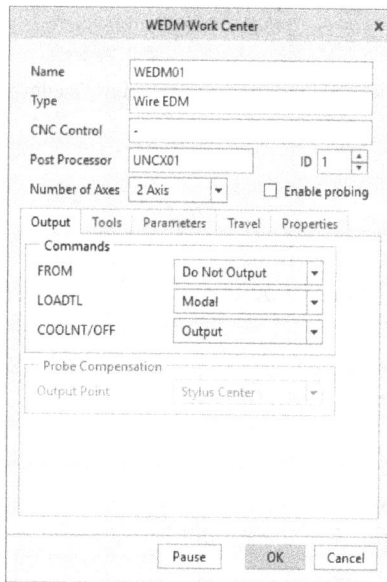

Fig. 9.54.   WEDM Work Center window.

(119) In the last window, select the **Tools** tab and then click on the
(**Tools…**) icon. In the **Tools Setup** window (Figure 9.55), provide
a proper name to the tool, e.g. **Wire**. Although the user is supposed
to enter a correct wire (cutter) diameter, such as **0.1**, **0.2**, **0.25**, or

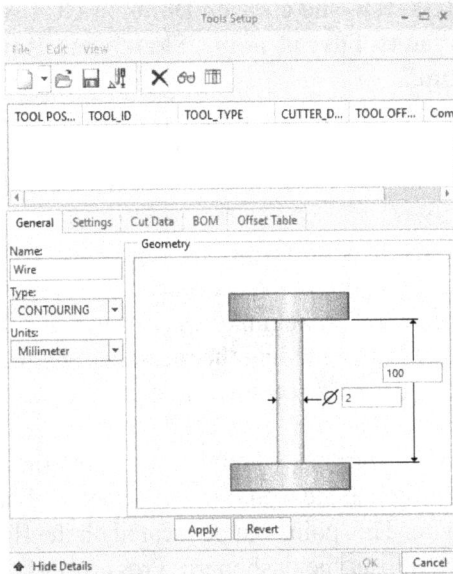

Fig. 9.55. Tools Setup window.

0.3 mm, for better visualisation of the simulation keep it larger, for example **2** or **3** mm. Change **Length** of the tool to be equal to the thickness of the material (workpiece) that is supposed to be cut. Click on **Apply** and close the **Tools Setup** window with **OK**.

(120) Close the **WEDM Work Center** window with **OK**.

(121) Using the **Clearance** tab, **Operation** dashboard, select an appropriate clearance plane above the part.

(122) Click on the green tick (✔) to close the **Operation** dashboard.

(123) Select the newly added **WireEDM** tab in the main ribbon and click on **Contour** icon (Contour) to open the **Menu Manager** window. Keep **Tool** and **Parameters** selected and click on **Done**. In the next **Tools Setup** window, click **OK** to confirm the use of the wire/tool that have been created earlier. In the next **Edit Parameters of Sequence** window, input the linear wire feed rate, **CUT_FEED** (highlighted in yellow) as **2** mm/min. Click **OK** to close the window.

(124) In the next window **Customize**, keep the default **Automatic cut** and click on **Insert**. In **Menu Manager** window, keep the defaults

Rough and **Sketch**, and click on **Done**. In **CUT ALONG** window, keep the defaults **Thread point, Sketch** and **Rough** checked and click on **Done**.

The system will expect inputs from the user in all these checked options. The first is the **DEFN POINT** menu for the **Thread point**.

It is good practice to look at the message area at the bottom of the screen in order to follow the input sequence and type of expected inputs specific to each command. A beginner can get lost in a long sequence of entries and can save time by reading the messages and responding accordingly. If you get lost, then **Cancel** the sequence and start again.

(125)   In wire EDM operations, when cutting an internal contour, a small starting hole inside the contour has to be drilled in advance to thread the wire. This point can be created on the fly by selecting the **Point** (ˣˣ Point) from the dashboard. Press (CTRL + Hold) and then click on the axis of the hole and the top surface to define the point position (intersection between the axis and plane) and **OK** to close. This point will be picked up as a threading point. Check this by clicking on **Show**. Click on **Done/Return** to close the last window.

(126)   In the next input window **SETUP SK PLN**, where the **Setup New** is pre-selected, the user is expected to sketch the contour. Click on the hole surface of the workpiece (normal to the hole axis) as a sketching plane. The next menu expects the sketching plane orientation. Select **Top** from the menu and then click on any wall perpendicular to the sketching plane as orientation reference.

(127)   In the next window **References**, select references for dimensioning, e.g. two perpendicular planes or edges. Click on **Solve** and, if the status is **Fully placed**, then **Close** the **References** window.

(128)   The Sketcher opens. Select a drawing tool (line chain, arc, circle, etc.) and draw the contour to be cut. If the cutting curve is an existing feature, as the hole in this case, then it is better to use the **Project** icon (▢) and copy the existing hole edge as shown in Figure 9.56.

(129)   If the contour is done, close the Sketcher with the green tick (✔).

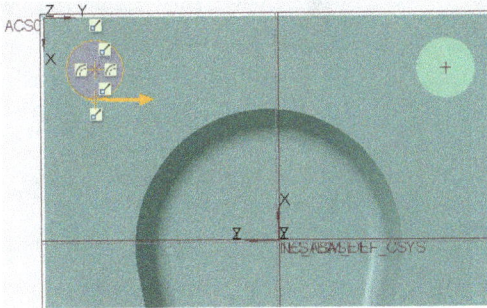

Fig. 9.56.   Cut contour as an existing hole.

Fig. 9.57.   Play Path simulation.

(130) From the **INT CUT** menu (**Menu Manager** window), test the cut using the **Play Cut** option. If the result is OK, close the last window with **Done Cut**. In **Follow Cut,** click **OK.** Close the **Customize** window with **OK**.

(131) From the **NC SEQUENCE** menu (**Menu Manager** window), select **Play Path — Screen Play** to run a simulation (Figure 9.57).

(132) Close the **Menu Manager** with **Done Seq** and save your model.

By looking again at the geometry of the reference part, speeds and feeds, processing steps and other parameters, the user could optimise the production processes.

Congratulations! You have completed the first steps toward realising manufactured products through the use of CAD/CAM software.

Chapter 10

Creating Engineering Drawings

10.1 Introduction

This lesson will introduce you to the procedure of creating engineering drawings directly from the 3D models.

Perhaps the reader has already learned how to draw an engineering drawing by hand or by using a 2D drafting software such as AutoCAD™. The main advantage of a 3D CAD system, such as Creo™ Parametric, against a 2D drafting software, is the higher level of automation. Within a 2D drafting system, the designer draws orthographic views by sketching lines, arcs, circles and other 2D entities, similar to hand drawing but more accurately. The 3D CAD software automatically generates all drawing views, projections and sections by referring to the actual 3D model geometry. The drawing will update automatically following the associated model modifications. In other words, the 3D CAD system will perform most of the technical and time-consuming work that a draftsman normally does. Despite all the automation, the designer must control the drawing workflow, specifying what views, projections, sections, etc. need to be created. Also, appropriate annotations such as dimensions, tolerances, surface finish, engineering notes, and other manufacturing data need to be placed manually in the drawing.

Aims:
- To introduce Creo <u>Drawing mode</u> and learn how to create a part and assembly engineering drawing directly from the reference model;

- To understand the basic steps in creating and modifying views, projections, sections and annotations.

Objectives:
- Learn how to create a new drawing file, select a template, format file, and sheet size;
- Learn how to generate views, orthographic projections, sections and detail views;
- Create dimensions, tolerances, notes, surface finish, and add the format frame and title block.

10.2 Drawing Workflow

Drawings are created by using Creo <u>Drawing mode</u>. Part or assembly drawings are generated following the same procedure. The difference is only in the reference model: a part (detail) drawing is associated with a part model while an assembly drawing is associated with an assembly model. Also, according to engineering drawing standards (ISO, BS, ASME, etc.) and practices, the part drawing contains all annotations necessary for the component manufacturing, while the assembly drawing presents the bill of materials and how the parts are assembled together.

The relationship and association between part, assembly and drawing files (models) have been discussed in Chapter 2 and shown in Figure 2.7.

The <u>Drawing mode</u> workflow consists of the following main steps:

- Start a new drawing file;
- Generate the main view and define orientation;
- Create orthographic projections, cross-sections, detail views;
- Add the following drawing annotations necessary for manufacturing:

  All dimensions;
  Manufacturing datum(s) such as planes and axes;
  Dimensional and geometric tolerances;
  Surface finish;
  Notes;
  Format and title block.

The first three steps are straightforward, and Creo automatically creates all views, projections, and sections. Adding annotations is not automatic, and it is the most time-consuming task. Referring to the design require-ments, manufacturing capability and drawing standards, the designer should make decisions about:

- How to place dimensions and dimensional chains?
- What dimensional and geometric tolerances are needed?
- What surface finish is adequate for the design quality?
- What other requirements are needed on the drawing to ensure that the manufactured component possess the expected functionally and qual-ity with the lowest production cost? For instance, it would not be wise to introduce a higher accuracy and narrow tolerances on dimensions that are not functional or fine surface finish on surfaces that do not need to be machined. Remember that the tolerances and surface finish placed on the drawing requires a specific manufacturing process that ultimately affects the production cost.

10.3 Create a New Drawing

A drawing is generated as a series of commands (features) that are initi-ated from the Drawing mode ribbon and recorded in the Drawing Tree. Every feature has a specific sequence of inputs, selected by the user enti-ties, such as view, datum plane, line, axis, point, etc. If the sequence is disturbed by a wrong input, then the command is unable to finish success-fully. The user should keep an eye on the Message area at the bottom of the Graphics window and follow the prompts.

The example below demonstrates how to create a detail (part) draw-ing of the SPINDLE.PRT following the drawing workflow.

(1) **Start** Creo (unless it is running already).
(2) Set up a Working Directory. Click (LMB) on **File > Select Working Directory** > C:\USER\CREO_PRACTICE.
(3) Click on **File > Open** and open the SPINDLE.PRT model created in Chapter 2.
(4) Start a new drawing. Click on **File > New**, or click on the **New** ( ) icon from the Quick Access Toolbar.
(5) In the **New** window (Figure 10.1), select **Drawing** as **Type**, enter SPINDLE_DRW as a name in the **Name** slot. Remove the tick mark for **Use default template** (unless it is clear what is the default tem-plate) and then select **OK** (OK) icon to accept.

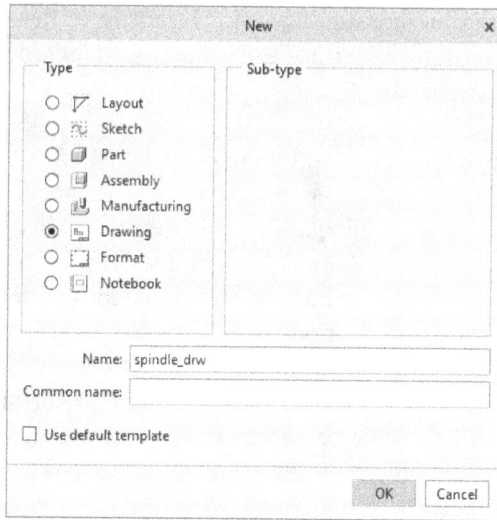

Fig. 10.1.   New drawing model.

(6) The **New Drawing** window opens (Figure 10.2). In the **Default Model** slot, the user inputs the reference model name, i.e. SPINDLE. PRT, which will be associated with this drawing file. The spindle part has been opened, that is why its name appears as default. Use the **Browse** icon to select another reference model (part or assembly). If the **Default Model** slot is empty, the drawing will have no association with any 3D model.

(7) In the **Specify Template** area, select **Empty** unless a template drawing file and/or a format files are available. In the **Orientation** area, keep the **Landscape** depressed. In the **Standard Size** slot, click on the down arrow to open the menu and select a suitable format size, i.e. **A3** size (a standard BS ISO engineering drawing size).

(8) Check all inputs and click **OK** (OK) to continue.

(9) The Drawing mode window opens, displaying the ribbon interface with commands and tools (Figure 10.3). Similar to Part mode, the drawing ribbon has tabs that switch the active ribbon. The **Layout** tab is active at the beginning of a drawing. Under each tab, the commands are arranged in groups. The Graphics area displays a rectangular frame with the selected A3 format size. The Drawing Tree and the Model Tree on the left side show the drawing and part names.

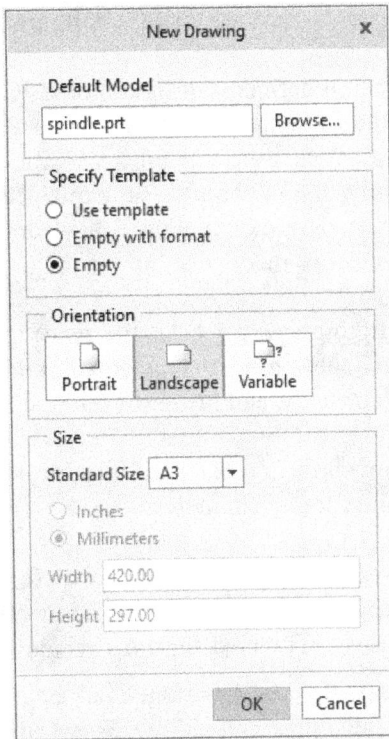

Fig. 10.2.   New Drawing window.

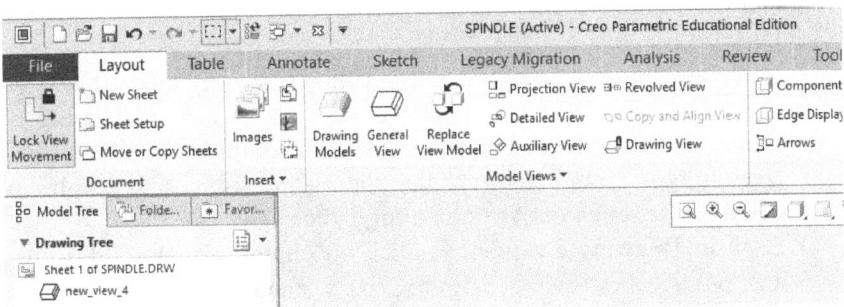

Fig. 10.3.   A fragment of the drawing ribbon (Layout tab).

10.4 Set Up the Drawing Configuration

The default drawing configuration is based on the American engineering drawing standards ASME Y14. The user can modify the default parameters and create a custom-made configuration for a specific application.

There are several predefined configurations, based on some main national and international standards, available in the **Drawing Setup Directory**. A good practice is to set up a specific standard automatically by reconfiguring the software, as explained in Section 1.14, Chapter 1. Instead, the user can specify a new drawing configuration standard or load a custom-made one at the beginning of each **New Drawing** process.

The drawing configuration (a *.DTL file type) includes hundreds of parameters, and it is a daunting job to change all of them. However, some of them can be changed easily, and the new custom configuration is saved in the working directory.

The next steps show how to replace the current drawing configuration with another and change some parameters.

(10) Go to **File** > **Prepare** and click on the **Drawing Properties**. In the opened **Drawing Properties** window, find the **Detail Options** row and click on **change**. The next **Options** window shows the content of the current drawing configuration. Click on the **Open** icon ( ) at the top of the current window. Select the **Drawing Setup Directory** and from the file list, pick and open the ISO.DTL. This file will replace the default parameters with those corresponding to the ISO technical drawing standards. Select **Close** to close **Detail Options**.

(11) In the ISO standard, the default projection type is 1st angle. The user can modify it to 3rd angle. To do this, go to **File** > **Prepare** > **Drawing Properties** and open the **Options** window. Find the **projection_type** parameter, change the value from **first_angle** to **third_angle**, click on **Apply/Change**, then **OK**.

(12) Back in **Drawing Properties**, click on **change** for the **Tolerance** and modify the tolerance to **ISO/DIN** standard. Click on **Done/ Return** and then **Close**.

Important: The drawing configuration is a property of the drawing file In order to use a custom-made configuration, it should be saved (and then opened) in a separate (*.DTL) file from the **Options** window.

10.5 Creating the Main View and Orientation

The main (general) view is the first feature of a drawing. It includes orientation, scale, display style and other view properties. All subsequent projection and/or section views are dependent on the main view. If the main one is deleted, then all views linked to it will fail to regenerate. The main view should be oriented to display most of the product details.

(13) From the **Layout** tab, click on the **General View** icon (_{General View}), select **DEFAULT ALL** (in **Select Combined State window**) and then **OK**.

(14) Look at the message area and notice that the system expects a <u>centre point for the drawing view</u> to be selected. Click on any point in the Graphics area to create the main view. The associated part appears in default orientation together with the **Drawing View** dialogue window shown in Figure 10.4.

The **Drawing View** controls the view properties listed in the **Categories** area as follows:

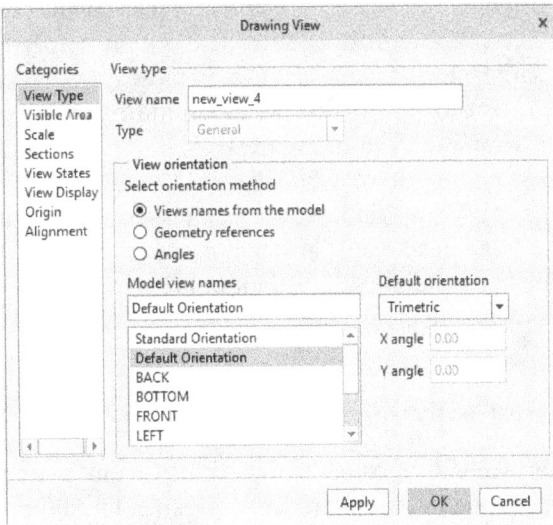

Fig. 10.4.   Drawing View properties control.

- **View Type** — Controls the view orientation. Some orientations are defined within the part model, i.e. **BACK, BOTTOM, FRONT**, and can be selected directly. If they are unsuitable, then switch the orientation method to **Geometry references** and provide inputs (datum plane or flat surface) in **Reference 1** and **Reference 2**. This is similar to selecting references for the **Sketch** command. Test the orientation by clicking on the **Apply** icon.
- **Visible Area** — By default, it is a **Full View**, but it could be switched to **Halve View**, **Partial View** and **Broken View**.
- **Scale** — Controls the view scale, i.e. the view size in relation to the model. In order to reduce or enlarge the view size, switch to **Custom scale** and specify the scale ratio. For example, a ratio of **2** corresponds to 2:1 scale, while a ratio of **0.5** corresponds to a 1:2 scale.
- **Sections** — Creates 2D cross-section for a given view.
- **View States** — Controls the exploded view in an assembly drawing.
- **View Display** — Controls the display style of a view. Normally, it should be set to **No Hidden** for a line drawing.

(15) Continuing from the previous step, in the **Drawing View**, set up the following: **View Type** — select **BOTTOM** or another suitable pre-defined orientation; **Scale** — enter **2**; **View Display** — select **No Hidden**. Keep all the rest as default. Click on **Apply** to test. If the result is satisfactory, then select **OK** to close.

(16) Switch OFF the display of all datums (from **Datum Display Filters**). The main view is shown in Figure 10.5.

If the **Drawing View** window has been closed, to open it back select the view in the Graphics area. If selected properly, a green dash dot rectangle will appear around the selected view. Press the (RMB + Hold), and from the pull-down side menu select **Properties**.

10.6 Creating a Projection View

(17) Projection view is created automatically from a reference view. Select the main view as reference. A green colour border around the main view appears after the selection. Next, click on the **Projection View** icon (⬚ Projection View) from the **Model Views** group (**Layout** tab) to activate. Move the cursor to the Graphics area and notice a

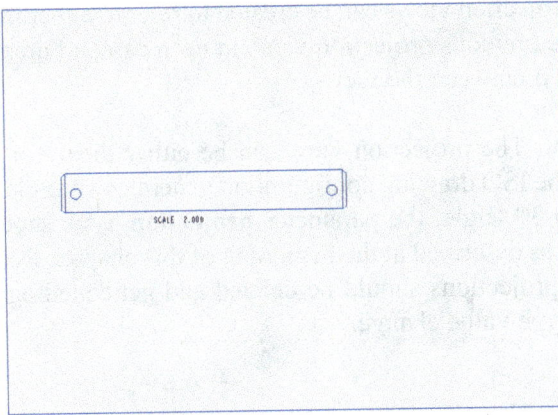

Fig. 10.5.   The main view in A3 format size.

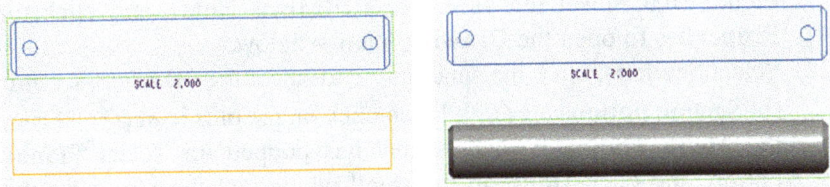

Fig. 10.6.   Projection view: preselection (left), and already placed (right).

yellow rectangle marking the projection view location (Figure 10.6, left). Click on a point to create the projection view (Figure 10.6, right).

(18) By default, the view movement is locked. To unlock it, select the main (<u>Parent</u>) view, and click on the **Lock View Movement** icon (⌐ Lock View Movement) at the top left area in the ribbon. After that, all views are unlocked to move.

(19) Notice that the projection view is shaded. To change its display style, select the view, press (RMB + Hold) to open the side menu and then select **Properties** to open the **Drawing View** control.

💡 Alternatively, change the value of **model_display_for_new_views** parameter to **no_hidden** (in **Detail Options**) as explained earlier.

(20) More projection views can be created to refer to either the main view or to the previous projection view. In each case, a <u>Parent–Child</u> link is created between the views.

ⓘ <u>Important:</u> The projection view can be either the 1st or 3rd angle of projection. The ISO drawing configuration generates 1st angle projections. To change to 3rd angle, the parameter **projection_type** should be set to **third_angle**, as described at the beginning of this chapter. Remember that any existing projections should be deleted and generated again after the **projection_type** value change.

10.7 Cross-Sections

(21) A cross-section is a projection view that has been converted into a section. To create a cross-section from the projection view in Figure 10.6, select the view, press (RMB + Hold), and click on **Properties** to open the **Drawing View** window.

(22) Select **Sections**, click the small circle to select the **2D cross-section** (in **Section options** area), and then click on the plus icon (✚). From the **Menu Manager** window that has popped up, select **Planar** option, and click **Done**. In the small window at the top, enter the section name, for example **A**, and click the OK green tick to confirm. Notice the prompts that appear in the message area at the bottom of the Graphics window. The current prompt asks the user to select a planar surface or datum plane in the corresponding view. (Switch ON the datum display.) Select the FRONT datum plane in the main view. If correctly selected, a green tick in front of (section) **A** is displayed. Click on **Apply** to preview the section (Figure 10.7).

(23) To display the section arrows on the parent view, move the slider (in **Drawing View**) to reveal all options from the right side. The last one is **Display Arrows**. Click inside the slot to activate and pick the main view where the section is perpendicular. Click on **Apply** to display the arrows and then select **OK** to close.

(24) Create a second projection view on the right side from the main view.

💡 There are many options in the Drawing View dialogue window that control the views and sections. Explore these options when you become more confident with the basics.

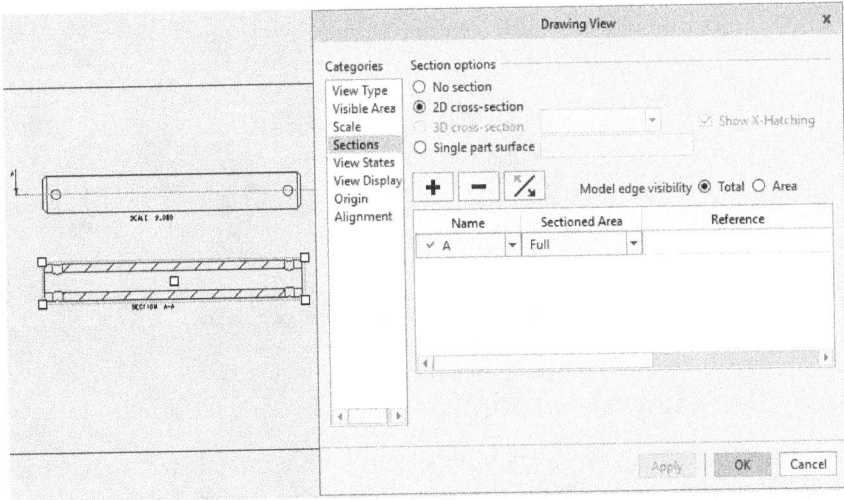

Fig. 10.7. A section created.

10.8 Creating a Detailed View

A detailed view is a fragment of a view or section containing small details, shown separately in a larger scale. The detailed views can be generated with the **Detailed View** command (**Model Views** group, **Layout** tab).

(25) Continue the previous drawing creating a detailed view for the spindle end in the cross-section.

(26) Select the **Detailed View** icon (Detailed View) from the **Model Views** group, **Layout** tab. Notice the prompt in the message area asking for a centre point. Zoom in to the spindle left end area and click on a point. Following the new prompt, click on another point to start sketching a spline around the area of interest. Click on at least 3-4 points surrounding the area (see Figure 10.8, left) and then press MMB to stop sketching (Figure 10.8, middle). The next prompt is to select a centre point for the view itself. Move the cursor to an empty area of the drawing and click on a point. The detailed view will be created in a larger scale, 4:1 (Figure 10.8, right). The scale and name can be changed in the same way as all other views, i.e. modifying the view properties in the **Drawing View** control window.

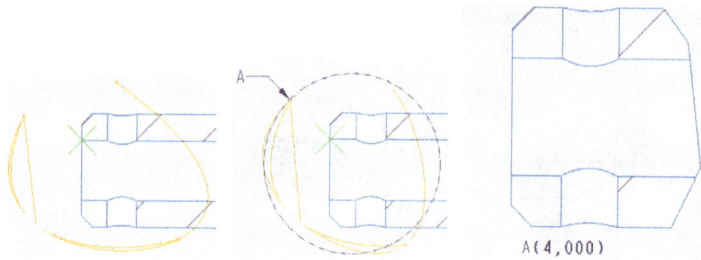

Fig. 10.8.   Stages of the detailed view generation.

10.9 Drawing Annotations

The drawing annotations are dimensions, dimensional and geometric tolerances, reference datums, surface finish and notes. All views and annotations update automatically when the reference model is modified. Commands that create annotations can be invoked from the **Annotate** tab.

10.9.1 *Creating dimensions automatically*

Since the reference model has been created, all current dimensions that define the shape are stored as feature parameters. Their values can be displayed automatically as dimensions on the drawing views using the **Show Model Annotations** command from the **Annotate** tab. This command allows all existing dimensions to be shown on the views.

(27)   Select the **Annotate** tab and click on the **Show Model Annotations** icon (Show Model Annotations) to open the control window. Notice that the first tab on the left **Show model dimensions** is active, and the **Type** is set to **All**. The user can select a feature and reveal its dimensions. Move the mouse cursor to the main view, slowly hover the cursor to preselect a feature and display its label, and click on the Extrude feature. The length and two diameters appear in the view and in the **Show Model Annotations** window. Select those that you would like to keep and click on **Apply** (Figure 10.9). Next, click on the Hole feature in the detailed view and the chamfer in the main view to show their dimensions. Select **Apply** after each feature.

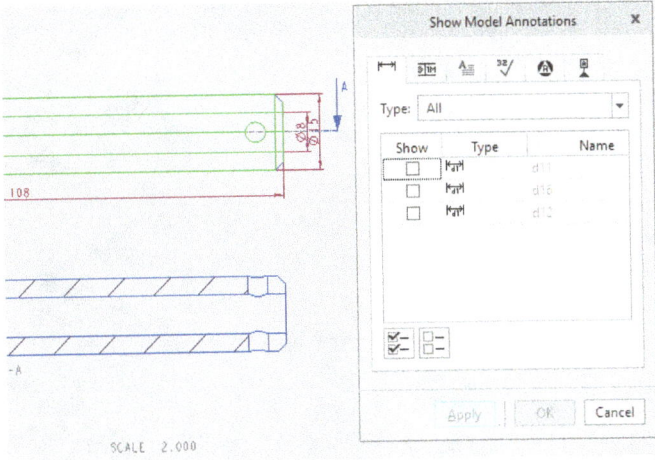

Fig. 10.9.   Creating dimensions automatically using 'Show Model Annotations'.

(28) Do not close the window yet and select the last tab (on the right) **Show the model datums**. Select the extrude feature in all views to reveal the axes, click on the axes in all views to show them, and then **Apply**.
Select **OK** or **Cancel** to close the window.

(29) Usually, automatically placed dimensions look messy. To tidy and separate them, select all dimensions by pressing (LMB + Hold), drag the mouse to draw a window around dimensions, and release the LMB to stop. Next, click on the **Cleanup Dimensions**, **Edit** group. In the **Cleanup Dimensions** window, adjust the **Offset** and **Increment** values and press **Apply**.

(30) Tidy the drawing by moving the views, dimensions, axes and text. To move an item, select it first, position the cursor on the item handle (a small white square, or a cross), press (LMB + Hold), drag the item to the new location and release the LMB. The drawing is shown in Figure 10.10.

10.9.2 *Creating additional dimensions*

Creating dimensions automatically works only with native Creo models because the **Show Model Annotations** command uses the values of the feature parameters. An imported model from another 3D CAD system

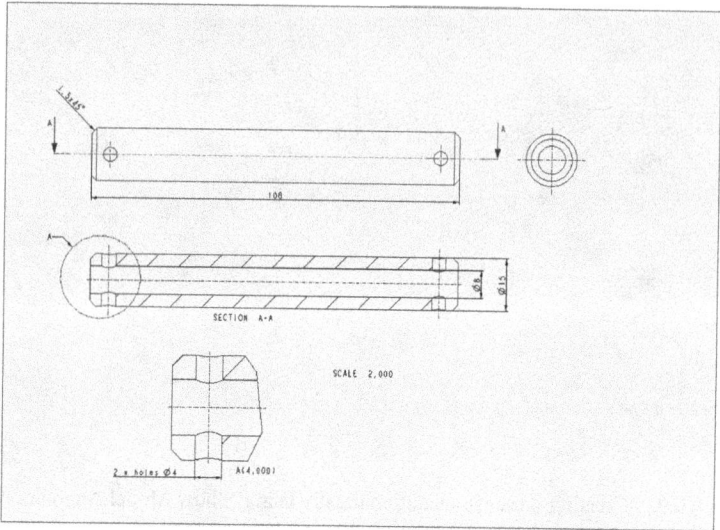

Fig. 10.10.   The spindle drawing with dimensions and axes.

(i.e. STEP or IGES file) cannot be dimensioned this way. In addition, some important dimensions might be missing because they do not have a suitable or corresponding feature parameter. In such cases, the **Dimension** command (**Annotate** tab) is employed to create dimensions directly on the drawing views.

In the spindle drawing, the distance between the two **4** mm holes is missing. The next step demonstrates how to add this dimension.

(31) Select the **Annotate** tab, and click on the **Dimension** icon (Dimension) to activate. The command expects two entities (lines, axis, points, arc) to measure a distance, or a single entity (arc, circle) to measure a radius or diameter. Click on the left **4** mm circle or its vertical centreline, press (CTRL + Hold), then move the cursor and click on the left circle (or centreline). A ghost dimension appears if the selection was successful (Figure 10.11, left). Move the cursor and MMB click to place the <u>distance</u>. Press MMB to cancel (stop) or carry on creating more dimensions.

(32) To create a <u>radius,</u> click (LMB) on an arc or circle, move the cursor and preview a ghost radius, and then click the MMB to place the <u>radius</u>. To create a <u>diameter,</u> click on a circle, a ghost radius appears,

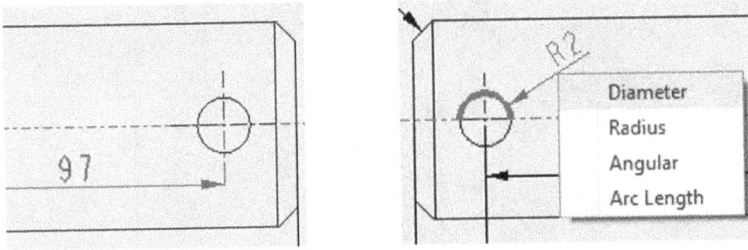

Fig. 10.11. Creating additional dimensions.

then click (RMB + Hold) to reveal the submenu, scroll and release the RMB on **Diameter** (Figure 10.11, right), and click MMB to place the diameter. Practice this step on the current drawing.

10.9.3 *Dimension editing*

Dimensions can be cleaned up automatically using **Cleanup Dimensions**, from the **Edit** group.

A dimension can be moved to a new location within a view by selecting it and then dragging it to the location. To delete a dimension, select it and press the DELETE key. Draw a window around several dimensions for multiple selections and press DELETE to delete them simultaneously.

Additional editing commands are available by selecting a dimension (or several dimensions) and then pressing (RMB + Hold) key to open the side menu as shown in Figure 10.12.

The most common editing commands in this menu are as follows:

- **Delete** — Deletes the selected dimensions;
- **Move to View** — Moves selected dimensions to a selected view;
- **Flip Arrows** — Flips dimension arrows;
- **Modify Nominal Value** — Modifies the dimension value and also the feature parameter value in the part model. After the change, the user should **Regenerate** the model to update the geometry. This is a reverse associativity that allows the reference model to be modified from the drawing. This functionality allows the user to make minor modifications and correct mistakes without opening the part model. However, it should be used carefully with some main features, positioned at the Model Tree top.

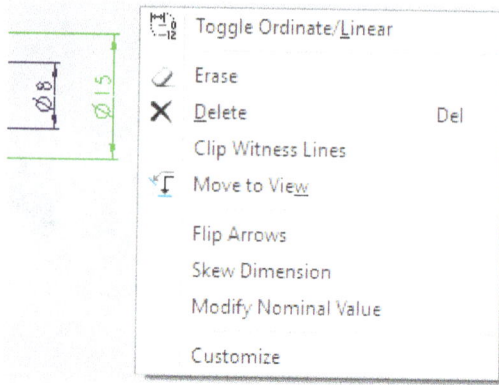

Fig. 10.12.   Dimension editing menu.

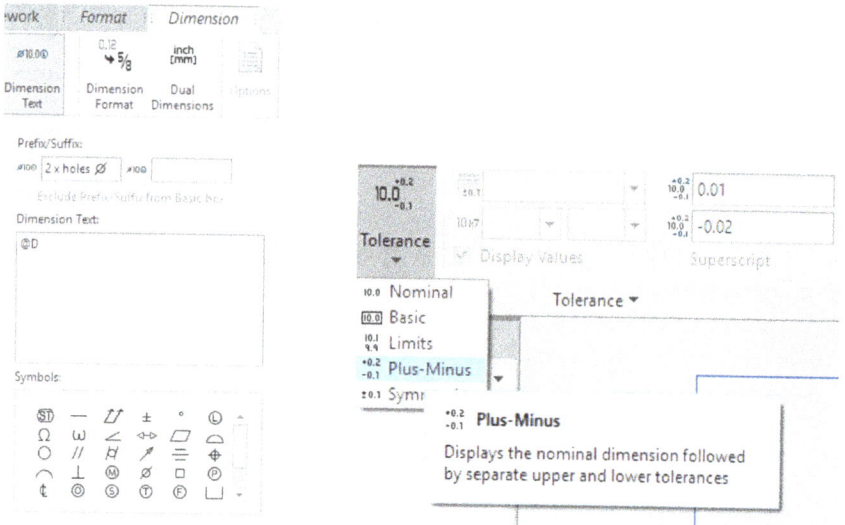

Fig. 10.13.   Dimension editing — Dimension menu (left), Tolerance menu (right).

10.9.4 *Dimension and format tabs*

The <u>dimension text</u> can be edited using the menus from two additional tabs *Format* and *Dimension*. They appear at the right end of the drawing ribbon, when a dimension is selected.

The *Format* sub ribbon has tools to apply different text styles, fonts, colour, alignments, etc. to the dimension text (Figure 10.13, left).

The *Dimension* sub ribbon has various tools for dimension text modifications, with the following main ones:

- **Dimension Text** — A menu to add/remove text and symbols as **Prefix**, **Suffix**, and add text to the dimension value slot. To add or remove text, select a dimension, and in the *Dimension* sub ribbon, click on **Dimension Text** icon to open the editing menu (Figure 10.13, left). Notice the **Symbols** area with variety of special drawing symbols to provide additional annotations.
- **Tolerance** — A menu to create a dimensional tolerance for the nominal value in the selected dimension. To apply, select the dimension to open the *Dimension* tab. Click on the **Tolerance** down arrow (the one on the left side) and open the drop-down menu. Scroll down and select the tolerance type that is most appropriate in this case. The available tolerance types are **Nominal**, **Basic**, **Limits**, **Plus-Minus** and **Symmetric**. Enter the deviation values in the two available slots on the right (see Figure 10.13, right).
- **Value** — This menu modifies the nominal value in the drawing and in the part model (the same as **Modify Nominal Value**).

The use of the dimension editing menus and tools is quite intuitive, and this section covers only the main principles of their application. After learning the main tools, the user is strongly adviced to practice the use of the other tools.

10.9.5 *Dimensional and geometric tolerances*

Dimensional and geometric tolerances are an important part of any detailed drawing. It is not possible to manufacture any product to exact size. The tolerances define the upper and lower limits of dimensional or geometric variation that should be achieved in production.

The tolerances are assigned by the designer and define the limits of size that will ensure correct part functionality within an assembly. Often, the tolerances define the type of interface between parts or the class of fit, i.e. loose, intermediate of tight fit. The values of the tolerance limits depend on the dimension nominal and accuracy grade.

Normally, the tolerances are placed together with adequate surface finish. The greater the accuracy required by the designer, the narrower the tolerance band and fine surface finish.

The tolerances and surface finish requirements impose the use of specific manufacturing processes that ultimately affect the production time and cost. It is undesirable to use the same (high) accuracy and fine surface finish on all component surfaces. These should be reserved for functional dimensions only.

(i) Dimensional and geometric tolerances can be placed on the drawing automatically using **Show Model Annotations** (Section 10.9.1). This advanced method works if they are already specified in the 3D model. A simpler method is to specify the tolerances directly in the drawing, as shown in the next sections.

10.9.5.1 *Dimensional tolerances*

The designer can express dimensional tolerances on the drawing in one of the two forms: general and/or individual. General tolerances apply to those dimensions that appear in a nominal format, that is, without tolerances. They are displayed in a tolerance table, often in the title block. The individual tolerances are placed on the functional dimensions and usually specify a higher accuracy grade.

It is recommended to represent all dimensions on non-functional surfaces with their nominal values only and specify the general tolerance as a note or in the title block. Every dimension important for the design functionality, requesting higher accuracy grades and tighter tolerances, should be represented by its nominal value and two deviations ('Plus-Minus' or 'Symmetrical') rather than 'Limits'. For example, these are the dimensions for fits, accommodation of seals, bearings, etc.

(33) Continue the previous example and place tolerances on some dimensions. Switch the ribbon to the **Annotate** tab. Select the **15** mm diameter dimension. The ***Dimensions*** sub ribbon opens. Click on the **Tolerance** down arrow to open the drop-down menu with tolerance types (Figure 10.13, right). Scroll down and select the tolerance type **Plus-Minus**. Enter the deviation values **0.01** and **−0.02** (including the minus sign) in the two slots on the right as shown in Figure 10.13, right. The dimension in the drawing view will assume the specified tolerances (Figure 10.14, left).

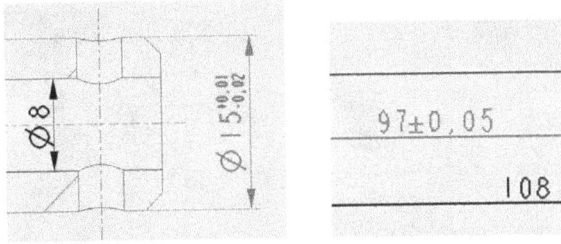

Fig. 10.14.   Geometric tolerances: 'Plus-Minus' (left), Symmetrical (right).

(34) Select the **97** mm dimension between the two holes. In the *Dimensions* sub ribbon, click on the **Tolerance** down arrow to open the drop-down menu, scroll down and select **Symmetrical**. Enter the **0.05** value in the available tolerance slot. The tolerance is added to the nominal dimension as shown in Figure 10.14, right.

The deviation values are defined and have to be taken from the corresponding Standard Systems of Limits and Fits and depend on the accuracy class (Standard Tolerance Grades).

10.9.5.2 *Geometric tolerances*

Geometric tolerancing takes part dimensioning further and considers the accuracy of the component shape. There are many types of geometric tolerances, such as cylindricity, flatness, parallelism, perpendicularity, position, circular run-out, etc. However, only those critical for the correct component functionality should be specified on the drawing.

A geometric tolerance can be attached to a dimension, datum (plane, axis), one or several edges and surface. Several tolerances can be stacked together and attached to the same dimension or datum. Also, they can be placed as a non-attached note in the drawing.

Before placing a specific geometric tolerance, the user should create a relevant tolerance reference and datum feature symbol.

(35) Select the **Annotate** tab in the ribbon, click on the **Datum Feature Symbol** icon (Datum Feature Symbol), select the spindle external edge or cylindrical surface as reference, move the cursor, and click the MMB to place the datum symbol as shown in Figure 10.15, left.

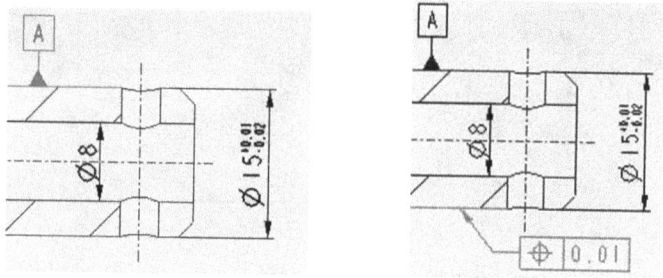

Fig. 10.15. Geometric tolerances: Datum feature symbol (left), "Runout" geometric
tolerance symbol (right).

⬙ A reference for the **Datum Feature Symbol** can be an edge, surface,
point, dimension or another geometric reference.

(36) Select the **Geometric Tolerance** icon (Geometric Tolerance) from the ribbon to
activate the tolerance placement. Select the external edge, move the
cursor and click MMB to create the tolerance. A default geometric
tolerance is placed (Figure 10.15, right). Also, the *Geometric
Tolerance* sub ribbon, controlling the tolerance, opens. Select the
Geometric Characteristics icon, scroll down and find the desired
tolerance type, i.e. **Runout**. Click in the primary datum field (the top
of the three slots in Figure 10.16), type and press ENTER. In the
tolerance value slot (the top of the two left slots in Figure 10.16)
type **0.02** and press ENTER. Click somewhere in the Graphics
window to finish and close the *Geometric Tolerance* sub ribbon.
(37) To edit a geometric tolerance, select the tolerance to open back the
previous sub ribbon and enter the modifications.

⬙ A reference for a **Geometric Tolerance** can be an edge, surface,
dimension or another geometric tolerance. If another tolerance is selected,
then the geometric tolerances are stacked together.

10.10 Creating Notes

Notes can be creating by selecting the **Note** icon (Note ▾) from the
Annotate tab ribbon.

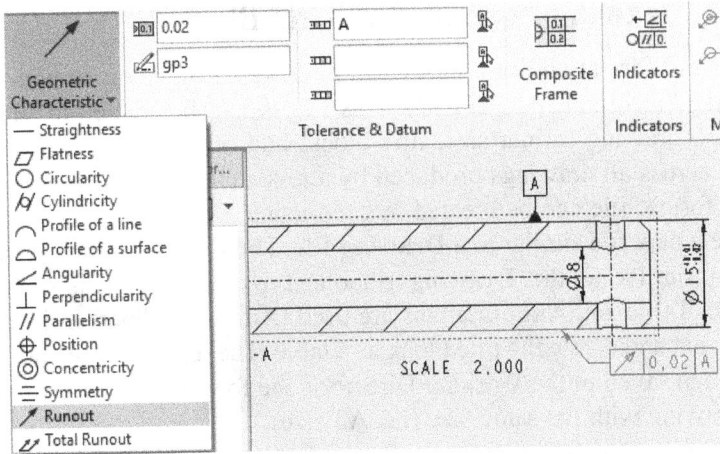

Fig. 10.16.   Geometric Characteristics menu (Geometric tolerance sub ribbon).

(38) Click on the down arrow to open the **Note** drop-down menu with a variety of note types (unattached, with a leader, on item, etc.) and then click on the desired note type to activate. Move the cursor in the Graphics area and select the note location.

(39) Type the note. Notice that the *Format* sub ribbon is open. The user can select various **Style**, **Text** (including drawing symbols) and **Format** options for the note.

10.11 Creating Surface Finish

(40) To indicate a surface finish, select the **Annotate** tab and then click on the **Surface Finish** icon (Surface Finish). In the **Open** window, the user can select what type of surface finish symbols to retrieve from the systems folder: generic, machined or unmachined. Double click to open one of the folders. Further, the user has two choices: **no_value** or **standard** (with value) option.

(41) In this case, select the machined folder and then click on **standard** 1. The **Surface Finish** dialogue window opens. Switch the **Placement** to **Normal to Entity** type. Place the symbol by LMB click on a surface. To finish, click the MMB. Continue placing more surface finish symbols if necessary. For each symbol (surface), enter the appropriate surface finish value in the **Variable Text** tab. Close the **Surface Finish** window with the **OK**.

10.12 Drawing Format and Title Block

A drawing format and title block can be inserted by using an additional file called a format file (*.FRM). This file should be created in advance for each drawing format size, title block, and other details that usually repeat across all drawings produced by a given engineering company.

A format file can be created in Creo using **File > New > Format** to initiate a new file similar to a **Drawing** file. The **Format** ribbon interface is also similar to the **Drawing** ribbon. Typically, the tools from the **Sketch, Table** and **Annotate** tabs are used to draw the lines of the drawing format and create the title block as a table. The text is created as notes.

When saved in the Working Directory, the format file can be inserted in a drawing with the same size (i.e. A3 size).

The procedure to insert a new format file or to replace an existing one (attached to the drawing) is as follows:

(42) Select the **Layout** tab and deselect any view. Move the cursor to an empty (no view) area of the drawing sheet, press (RMB + Hold) to

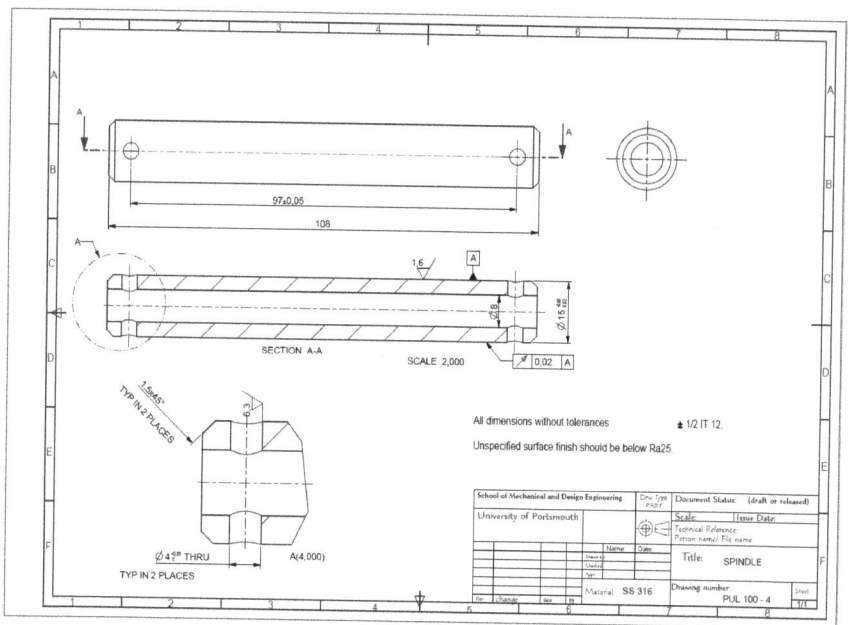

Fig. 10.17. The SPINDLE drawing.

open the side menu and select **Sheet Setup**. (Alternatively, click on the (⬚ Sheet Setup) icon under the **Layout** tab.)

(43) In the **Sheet Setup** window, click on the slot with the current format (i.e. A3 Size) under the **Format** to activate. Next, click on the black down arrow to open the drop-down list of available formats.

(44) Click on the last item **Browse** to select a new format size from those located in the System Formats or from the Working Directory.

(45) The selected format containing the frame and title block is attached to the drawing as shown in Figure 10.17.

ⓘ A single drawing file (*.DRW) can contain several drawing sheets with the same or different format. To create another sheet, click on the (+) icon at the bottom left corner of the current sheet. Click (RMB + Hold) on the **Sheet 1** icon (Sheet 1) to open a menu with more sheet commands.

10.13 Printing a Drawing

A drawing can be printed from the **Print** setup ribbon. Click on **File >**

Print > Print (🖨) to open the print setup and have full control on all printing options. However, a quick and easy way to print a drawing is by creating a PDF file first. To do this, select **File > Save As > Quick Export (*.PDF)**. The generated *.PDF file can be printed on any printer connected to the PC or network.

Congratulations! You have completed the last chapter and learned how to create a drawing directly from a 3D solid model.

Index